A Neolithic Universe

(Expansion of Stonehenge: Solving the Neolithic Universe, 2nd edition)

By Jonathan Morris

FRONT MATTER

Copyright © Jonathan M. Morris
© MMXIII, MMXX

All rights reserved. All images, photographs and other image types copyright of the author or others. Except for the quotation of brief passages in criticism, no part of this publication may be reproduced, stored in a retrieval system, or transmitted, in any form or by any means, electronic, mechanical, photocopying, recording or otherwise, without the prior permission of the publishers. The Author asserts his moral right to be identified as the author of this work. Definitions and meanings of patents, designs and copyright described above should be understood in accordance with the laws of the United Kingdom.

www.aneolithicuniverse.com

Original First edition (electronic) published November 11, 2012, Eastbourne, UK. Published by Handow (Hanwell & Dowling) Publishing using the Createspace and Kindle platforms under ISBN numbers 978-1481876353 and/or 148187635X. Expanded second edition first published at the Autumn Equinox, 2013, Using ISBN numbers 13: 978-1492736882 and/or 1492736880

This third edition published by Handow (Hanwell & Dowling) Publishing,
October 2020
ISBN-13: 978-0-9568617-3-3

ACKNOWLEDGEMENTS

For Inspiration
I would like to thank the following authors:
Aubrey Burl, Rodney Castleden, Christopher Chippindale, Rosamund M J Cleal, Timothy Darvill, FE Halliday, Gerald S Hawkins, Rosemary Hill, Anthony Johnson, Jean-Pierre Mohen, R Montague, Mike Parker Pearson, Mark Pendergrast, Mike Pitts, F Pryor, Julian Richards, J M Roberts, Lewis Spence & K E Walker.
And, especially, Wikipedia.

In particular, this book would not have been possible without Anthony Johnson's book: 'Solving Stonehenge' and the English Heritage book 'Stonehenge in its Landscape'.

Engineering and solar concepts in The Broken Stone were expanded upon in various discussions within the Megalithic Portal Forum, whose contributors kindly offered advice and criticism. I would also like to thank George Currie and Neil Wiseman for their input into Chapter 2, for general comments, Rob Ixer for his comments on metals, Rune for comments on Newgrange, Simon Banton for comments on the A posts, Andrew Davies for information on the area around Pen y Fan and Timothy Darvill, Sue Greaney and Simon Charlesworth for additional comment.

I am in debt to NASA, Neil Wiseman, Terence Meaden and CIBSE for kindly creating and/or supplying some of the images.

RESERVATIONS

Archaeological references have been included only for the purpose of allowing accelerated checks of dimensions and other data. Where archaeological references have been included, they have usually been chosen to be readily publicly accessible. However, a great lack of diversity of authorship was encountered in all of these types of reference. In addition, older works referred to within the text can be quite outdated and should be treated with caution.

FOREWORD

The features of the monuments described in this book, which include Stonehenge, are shown to be explainable using simple science and engineering principles. The book describes how those monuments could be the result of a search for knowledge; resulting in an early fundamental view of the Universe and the subsequent creation of Stonehenge itself.

The book contains workable explanations for a number of those monuments. Each of these explanations is logically linked to explanations for other monuments; and this set of explanations, as a whole, leads forward to an explanation for the need to build Stonehenge itself.

In this book, none of the explanations contain conspiracies, ritual, aliens, anything else other than humans, psychic influences, or any reliance on other unknowable factors. Every explanation can be checked using ordinary materials. Some of the methods are unusual, but they are not 'lost' or 'advanced' technology. All of the explanations can be checked by going to the sites referred to.

The two possibilities for these explanations are:

1) The author has stumbled across why some ancient monuments in Northern Europe were constructed: each explanation linked to other explanations in the series; all of which are new, workable and, simultaneously, are all the result of the same logical sequence.

2) The author is capable of making up alternative explanations for past events known to be linked: each explanation linked to other explanations in the series; all of which are new, workable and, simultaneously, are all the result of the same logical sequence.

If you can not find a technical fault with some, or all, of the explanations which follow, which of the above is more likely?

The images of Stonehenge in the following text were produced using a three-dimensional computer model of a newly invented renewable energy device. This device was later found to replicate the stones at Stonehenge as they may have been when first constructed. In addition to working as a solar concentrator, this new invention can also be used to create a miniature version of the Universe assuming that the world is fixed at the centre.

Stonehenge's plan layout will be shown to be the same as an idealised geocentric description of our Universe. Its inner stone monument is demonstrated to be capable of producing a spectacular public display of solar movement. The arrangement of this system is shown to be based on a simple method of tracking celestial objects.

Other monuments will be shown to be capable of being used to find out what type of Universe would be thought to exist. This based on evidence found by using those monuments. The early view of the Universe that would be created will be shown to be based on geocentric principles.

This description of the Universe is almost identical to the same ideas produced in Ancient Greece some two thousand years later; when Ptolemy, Plato, and Aristotle were to discover how their Universe worked.

However, Ancient Greece does not credit Grecians with this discovery. Instead, a mysterious and obscure Titan, known as Hyperion, is credited. One of the sons of Heaven and Earth, Titan was also the father of the dawn, the Sun, and the Moon. Little is known about Hyperion and there is little to no reference to Hyperion during the War of the Titans. Similar myths, telling of a great war, exist throughout Europe and the Near East: One generation of Gods, or perhaps tribes, or even ideas, opposing those in power to win out the day.

Whoever *or whatever* he was, Hyperion was perhaps too far away from Greece to be significantly involved in their war:

> *Of Hyperion we are told that he was the first to understand, by diligent attention and observation, the movement of both the sun and the moon and the other stars, and the seasons as well, in that they are caused by these bodies, and to make these facts known to others; and that for this reason he was called the father of these bodies, since he had begotten, so to speak, the speculation about them and their nature.'*
>
> Diodorus Siculus, Library of History 5.67.

Table of Contents

1: SKIES AND EARTH 1
At the edge of the known world 2
- An introduction to Stonehenge 2
- Stonehenge: Chapters 5 and 6 2
What is our World? 3
- The size of the world: Chapter 7 3
- That the Earth is a sphere: Chapter 8 3
- The connection to Stonehenge: Chapter 9 3
Motivation: Fear 4
- Fear of the Sun moving South: Chapter 10 4
- Describing the fear: Chapter 11 5
Folklore, the Makers' Mark and Two Mountains. 6
- Folklore: Chapter 12 6
- The Makers' Mark: Chapter 13 6
- Chapter 14 6
The Universe, coincidence and epilogue 6

2: AN INTRODUCTION TO STONEHENGE 7
The people 8
The pathways of the high ground 10
Stonehenge: The word 12
A history of discovery 14

3: A POTTED HISTORY AND THE THEORIES 18
A potted history of the monument's phases 19
- Potted history: Stage 1 (3000-2920 BC) 21
- Potted History: Stage 2 (2620-2480 BC) 23
- Potted history: Stage 3 (2480-2280 BC) 25
- Potted history: Stage 4 & 5 (2280-2020 BC and 1680-1520 BC) 26
The Stones: A potted summary 27
The axis of the monument 31
Theories 33

4: MIRROR, MIRROR 35
Extraction 35
The source of metal 36
Polishing and the reflectance of tin 38
Antiquity and Stonehenge 39

5: STONEHENGE AND THE HINGE 41
The times 42
Astronomy on a fixed world 43
The Hinge of the Heavens 51
Apparent Orbits and Station Stones 56
Accuracy of the design 58
A summary of Stonehenge's layout 60

6: A 'T' SHAPE DEVICE & THE COSMOS	61
The 'T' Rune in antiquity	62
The three season device	68
Pre-testing	75
The 'T' Rune	76
Decay	79
Summary	80
7: THE SIZE OF THE WORLD	81
History	82
The idea	83
Finding a hill's height	85
A modern experiment	92
Calculating the size of the world	93
Describing how to calculate the size of the world	94
Summary	97
8: NAYSAYERS TO A ROUND WORLD	98
The rotation of the Heavens	99
The Long Man of Wilmington	101
Two sticks and a hill's height	103
Two sticks at Wilmington	105
Sunrise and sunset	109
It's just here. Nowhere else	110
Summary	111
9: BEGINNINGS	112
Eryr and Feddau	115
Cwmcerwyn	117
Numbers and bases	120
Milk Hill, Tan Hill, Foel Feddau and Foel Eryr	123
Summary	125
10: FEAR (EARLY MONUMENTS)	126
Newgrange: The Sentinel	126
The fear	128
A method of testing the fear	133
A summary of the sentinels	137
Other chambers not aligned to solstice	138
11: EXPLAINING THE FEAR	140
An explanatory layout	142
Beginnings and endings	145
Summary	147
12: FOLKLORE	148
The Grail	148
Treasures of the Tuatha Dé Danann	151
The Druids	152
Drych Haul Cib Dâr	153
Summary	154

13: THE MAKERS' MARK 155
    Stone 3: Scottish Island groups: *158*
    Stone 4: British and Irish groups: *158*
    Stone 5: Continental European groups: *159*
   Summary 160
14: TWIN PEAKS 161
    The cliffs *171*
15: THE UNIVERSE 173
   Coincidences: Part 1 175
   Coincidences: Part 2 177
   Summary 185
16: EPILOGUE 187

APPENDIX A: The rotation of our planet 189
   Some technical stuff 191
   Apparent solar planes 192
APPENDIX B: Spherical solar collectors 195
   Background 196
   Theory 197
   Before Stonehenge 201

NOTES & REFERENCES 209
   References 227
INDEX 232
   List of illustrations 238

1: SKIES AND EARTH

This book describes, and shows the evidence for, the idea that some Neolithic monuments in Britain, Ireland, and the European Continent are the result of early research into the nature of our world.

Chapters two to four contain referenced background for the hypotheses. The fifth to thirteenth chapters look at different aspects of various monuments and then tie each of those in to the central ideas. Chapter five is central to those ideas, but it is possible to discard one or more of the following chapters (or sections) without affecting the general direction, and the overall body, of circumstantial evidence. For this reason, I have tried to keep each chapter as a 'self-contained' work.

Decide what you like and discard those parts you do not. If a set of individual chapters stands up to your eye, they can be read together (without needing the parts that you feel may just be coincidence).

For people living five thousand years ago in Britain and Ireland, at the edge of the known world, the skies above would have looked different to today. For over a thousand years, the North Galactic Pole would have passed overhead at a position which is almost directly above Southern Britain.[01] The Milky Way would have regularly retreated to the horizon as a result, giving some very dark night skies.

For reasons unknown, people may have become concerned that their environment might change and, as a result, their lands might become uninhabitable. This book describes the idea that some ancient monuments, in this part of the world, were invented out of necessity rather than for a 'ritual' purpose. For example, if the circular Stonehenge describes our world, then evidence should exist elsewhere showing how that knowledge was obtained. However, the book makes no claim that all ancient monuments were utilitarian rather than 'ritual'.

At the edge of the known world

In the centuries before the Stone circle was built at Stonehenge, many other stone circles, earthworks, and monuments were constructed. If some of those constructions had an original purpose, then others built later may be copies, or 'homages', to the original constructions.

For example, there are dozens of modern Stonehenge lookalikes spread throughout the world,[02] but only Stonehenge contains all of the unusual, and original, features. The people who built the recent 'Clonehenges' did not know what purpose Stonehenge served, so only included those features that may have seemed important to them.

Given that so many 'Clonehenges' have been made in just the last century or two, it would be difficult to know which Neolithic monuments are also lookalikes; homages made in the Neolithic. Some of those may be copies designed only to impress. However, those ancient structures whose construction would have needed vast effort have a much higher probability of also being the ones that have an original purpose (rather than just being constructed as a homage). None of the modern 'Clonehenges' took anywhere near as much effort to build as the original.

For this reason, this book looks largely at monuments that were both hard to build and have a large number of unique features. These monuments also tend to be the ones that are well known.

An introduction to Stonehenge

Chapters 2 to 4 look at what exists at Stonehenge, its sequence of construction, existing theories, and some notes on the early use of metal. The reason for looking at early metal will be explored in chapter 6.

Stonehenge: Chapters 5 and 6

The idea central to this book is that Stonehenge may have been intended to be the final stage of development of very early ideas about the Cosmos. If so, it would have shown a 'geocentric' understanding of that Cosmos.

Chapter 5 looks at how the layout of Stonehenge, including the sarsen ring and beyond, can be used to represent the Cosmos using a 'world fixed at the centre' (geocentric) philosophy. Chapter 6 looks at how the structure contained within the sarsen ring can also be used to demonstrate how a geocentric (fixed world) Cosmos works.

1: SKIES AND EARTH

What is our World?

It is sometimes said that the world can be seen to be curved by watching a ship pass over the horizon. But in Neolithic times, the boats could not have been big enough for people to have seen this effect.

Because the boat effect could not have been seen, the world may once have been thought to be flat or some form of curved disc. One way to find out which idea is true is to measure the angles down to the sea's horizon from hills of a known height. This is a very simple experiment which needs only a few sticks. Finding the height of a hill is also a simple task, but it requires a lot more effort.

If the world were flat, the angle down to the horizon could show how far away the edge of a 'disc-world' is. This can be found using the measured angle and the hill's height. If places are known to exist farther away than calculated, the world can not be a flat 'disc-world'. Other questions about the nature of the world can also be tested using this method.

At a large number of places across Britain, France, and Ireland, Neolithic monuments exist at the exact location which would allow those angles to be measured.

The size of the world: Chapter 7

Chapter 7 looks at one location, along the South Coast of Britain, which has an exceptionally high density of Neolithic monuments (the far eastern end of the South Downs). It shows that these locations, and the types of monument found, are all suited to measuring the size of the world.

That the Earth is a sphere: Chapter 8

Chapter 8 shows additional methods, monuments and locations which can also be used to confirm that the world is a sphere. This chapter gives examples of how this could be done and also shows where to find other locations at which those tests can be done (together with associated Neolithic monuments that exist at those locations).

The connection to Stonehenge: Chapter 9

Chapter 9 looks at locations and monuments which might have started this process. It also shows how these locations (Avebury and Preseli) appear to be directly connected to Stonehenge and then describes why Stonehenge's layout appears to be directly connected to those processes.

Motivation: Fear

Our world appears to be fixed and the celestial sphere, above our heads, appears to revolve around the Earth's polar axis. Within that celestial globe, the Sun appears to move from south to north, over the course of six months, before returning south for winter. Even today, we describe the Sun as 'rising' in the morning and 'setting' in the evening. Unlike most other celestial bodies, the Sun's apparent orbital plane, as it moves around our world, changes over the seasons.

Until Galileo championed heliocentrism, it was commonly thought that the Earth was fixed and located at the centre of everything (geocentrism). In the areas surrounding the Mediterranean, Geocentric ideas can be traced to pre-Socratic society (before 500BC).[03] Those ideas may have existed earlier but, if so, no record of that knowledge exists.

The Moon also behaves strangely if you believe you are on a world fixed at the centre of everything. Whilst the Sun appears to change the plane of its orbit annually, the Moon appears to change in a monthly cycle (between 'lunistices'). Crucially, the Moon also changes how far it appears to go, in that monthly back & forth movement, between what is known as lunar 'standstills'. That second cycle of movement lasts for 18.6 years.

Fear of the Sun moving South: Chapter 10

Over the period when the Sun seems to move to the South (winter), the land becomes cold. When it returns to an apparent orbit over the North, the northern lands become warm (summer). With the change of seasons, the same pattern of hot/cold weather repeats every year.

This effect is due to the tilt of the Earth's axis as we revolve around the Sun. However, Neolithic people would not have known that our planet is a speck of dust in a vast Universe: to them, our world would seem fixed. Both the Sun and the Moon would seem to exist within, but not be anchored to, the celestial sphere and its fixed stars.

In times of unusually cold weather, a huge concern might be that the Sun could have moved, or decided to move, towards the South. A series of cold winters combined with bad summers would reduce food supply and put stress on the community. Finding out how the heavens work could have been a priority for anyone living through that type of event.

If the Sun were capable of slowly moving to the South, the whole community would eventually have to move southwards.

Chapter 10 looks at methods of taking accurate measurements of the Sun's movement using nothing but stone and timber. One of the most accurate and efficient systems, using a light-box, appears to be the same as the structure which currently exists at a place known as Newgrange. The detail of this method, and why that matches what is found at that monument, is described in this chapter.

Newgrange seen from Knowth, Ireland

Describing the fear: Chapter 11

Chapter 11 describes how to draw a picture of how the Sun might start to change its cycle in a way that could be understood. That type of drawn diagram appears to exist in the layout of Avebury; and that layout could have been used to explain how people could find out if the Sun had started to change its cycle. The experiments needed to find that out can, or could, be done in places that have a structure similar to Newgrange.

The barrow of East Kennet, close to Avebury, appears to have had a similar structure to Newgrange. Its alignment is also towards the same solstice direction as found at Newgrange.

Another way of finding the minor and major standstills of the Moon is by using posts set in the ground. The monthly swaying cycle of the Moon is a little like the movement of a branch on a tree, where the whole tree also is swaying over a much longer cycle. From the point of view of a geocentric observer, the movement between the major, and minor, standstills of the moon are similar to the sway of a tree.

Chapter 11 looks at a system to find out how the moon sways between cycles. A series of timber post-holes were found close to the original entrance to the Stonehenge earthworks. Relative to whatever was at the centre of Stonehenge, this line of posts (known as the 'A Posts'[04]) could have been used to find the 'travel' of the Moon over the 18.6 year cycle between its 'lunistices'.

The chapter also shows that the archaeological remains produced by this method would be the same as were found to exist in the very early stages of Stonehenge. It would be fitting if a new monument was built to overwrite the history of the place which helped to start the fear.

Folklore, the Makers' Mark and Two Mountains.

Folklore: Chapter 12

If some, or all, of the events described in chapters 5 to 11 happened at a far distant time in the past, some trace of those events might be found in folklore. This chapter reviews whether or not folklore has any similarity to the events described. In particular, it focuses on the legends of The Grail, Treasures of the Tuatha Dé Danann, The Druids and Drych Haul Cib Dâr.

The Makers' Mark: Chapter 13

If some or all of the events described in chapters 5 to 12 happened, it is also possible that the Makers of Stonehenge made some sort of signature to identify themselves. This chapter reviews whether or not the marks found at Stonehenge could be used, as marks of identity, to describe who the builders were.

Chapter 14

This chapter looks at places not known to be associated with Stonehenge.

The Universe, coincidence and epilogue

The last two chapters summarise some of the coincidences.

2: AN INTRODUCTION TO STONEHENGE

Stonehenge is one of a thousand or so stone circles within the British Isles. Similar circles also exist in some parts of Northern Europe. But a henge is a circular structure, more usually an enclosure, and is quite peculiar to Britain.[01,02] Surprisingly, Stonehenge is not a henge.

Nevertheless, Stonehenge is a truly exceptional circle: it is one of the few monuments in Britain where megaliths were carved to shape.[03] Huge stone lintels, weighing 5-6 tons[04] and locked together with joints similar to those found in woodworking,[05] were placed onto massive 25 ton[04] sarsen stones to make a giant 30 metre (≈100') internally facing[06] ring of stone.

Inside the ring, even larger stones, weighing up to 40 tons and with lintels up to 16 tons,[07] were laid out in a horse-shoe pattern which points north-east in the direction of the summer solstice sunrise. The tallest of these stones would have been about five times higher than the average person.

Stonehenge: As seen from north-east

The people

Four and a half thousand years ago, the people were very much like the inhabitants of modern Britain: it is thought that many of the genes in the population, who currently live in the United Kingdom, are directly descended from the early hunter-gatherer groups who thrived, and moved ever northwards, as the glaciers of the last ice age retreated.[11]

But at Stonehenge, the people were not all from the local area. At nearby Boscombe Down, the burials are known to be of people who had come from Wales or the Lake District.[12]

Not only were the people not all local, but some were not even born in Britain. The richest burial of the time was of a man who died aged 35-45 years, and is known as the 'Amesbury Archer'. Analysis shows that he died some time between 2200-2400BC and came from Central Europe.[13] Copper, in one of his knives, is known to come from Spain. No-one knows what caused him to come to Britain, nor why his possessions came from so far away. This island, once part of mainland Europe, had become isolated from the continent some three thousand years earlier.

But the Amesbury Archer was only the start of this migration.[14] Something special was going on in Britain. Cross-channel trade was starting to develop and, in the later middle Bronze age, would expand on a much larger scale.[15] People were coming to this island from far and wide. At some other time long after, another great wave of people came from the Atlantic seaboard[16] to settle in Ireland, Scotland and the west of England.

Though life was hard compared to modern times, the population was small and had access to ample natural resources. Clothing was made of fur, leather and vegetable fibres such as flax (linen).[17] Although simply made, it is possible that some clothes of that era would not look out of place today.

Surrounding Stonehenge, a partially wooded landscape, with small arable plots and large areas of grass, provided both grain for the people and grazing for livestock.[18] Grown during the warm summers, grain could be stored in granaries[19] for the winters; which were slightly colder than the winters of today. This was very much a communal lifestyle.[20]

2: AN INTRODUCTION TO STONEHENGE

Near to Stonehenge, the town of Durrington was vast by neolithic standards. It covered 17 hectares (42 acres) and was filled with rows of small square houses about 5 metres (16') wide.[21] Much of the activity in this town appears to have taken place over a very short period, leading archaeologists to think that construction may have lasted only decades rather than centuries.[22]

At this town, red deer antlers were used for digging and construction on "a truly grand scale":[23] At least 400 discarded antler picks have been discovered. This work began between 2525-2470 BC and ended some time between 2480-2440 BC.[24] But radio carbon dates show that the town continued to be in use for quite a while longer.[22]

To the east, the River Avon flowed much deeper and wider than today,[25] providing a ready source of water for the town. It is possible that river craft also sailed these waters. Rope, made from honeysuckle or the stripped bark of the lime tree, would have been readily available and longboats & frail coracle-like craft had been around for some 3000 years.

The archaeologist and writer Francis Pryor has noted that *"Britain is alone in Northern and Atlantic Europe in having produced several examples of Earlier Bronze age plank built boats".*[26] Even at the earliest stages, the people of the Northern islands of Europe appear to have been sea-faring.

In some ways, the people of the area were different. Some Wessex skulls have been found to be quite child-like: men may have been quite similar in appearance to women and their facial features refined and dainty. With noses, for instance, small and turned up,[27] these people appear to have been young and of a slight and slender build.[28]

But near Stonehenge, they were not relying solely on stored grain to survive. Instead, the people were feasting in mid-winter (and also possibly mid-summer[29]) on young roasted and stewed pork[30] brought here from afar.[31] We know of the winter timing because the pig teeth came from domestic animals born in the spring; nine months before the winter slaughter.[32]

This was a party. It was a big party and it was filled with young men and women. Whatever was going on, it had become so well known that people from all over Europe appear to have arrived to join in.

The pathways of the high ground

Though we know little of the lowland tracks, at high level it is easier to see where you are going; so these routes probably formed one of the easiest method of early travel and discovery. The Great Ridgeway, (also known as the Icknield Way), is an ancient track-way which runs from Norfolk across the Chilterns to Avebury, where it joins with a track-way from the Cotswolds before turning down towards, and then around, Stonehenge. Another track-way, the Harrow or 'Old Way' (and now sometimes known as the Pilgrim's Way), runs from the farthest point of Kent and then across the North Downs, again to Stonehenge.

The Harrow way is noted in old texts as having a branch which runs up to Avebury and down to Winchester.[41] The main route continues on, along or parallel to what is now the A303, and eventually meets up with the Great Ridgeway before running down to Dorset. In this vicinity, many other tracks meet and lead down to Old Sarum (Salisbury), the West and its tin mining country:

Composite map of routes
Map compiled from various references: (See notes: 42-44)[42-44]

2: AN INTRODUCTION TO STONEHENGE

The South Downs track-way, known to be some eight thousand years old,[45] runs upwards from Eastbourne alongside a dense network of neolithic earthworks close to Combe Hill.[46] It then turns westwards through areas only fully cleared of trees in the later Bronze Age.[47] From here, it continues along the high ridges of the South Downs before turning north at Butser Hill (in Hampshire) to continue on to Winchester. Beyond Winchester, it joins other track-ways from the west of England, where some paths are known to be the oldest man-made roads in Europe.[48]

From Christchurch, on the south coast and directly south of Stonehenge, another track[49] leads to the same path-way network and then continues on towards Stonehenge. The map shown previously is a compilation of probable main routes which have been compiled from various sources.[42-44] But no doubt there were many other tracks; far more than can be shown at this scale.

In the old visitor's car park at Stonehenge, three pits, probably containing posts, have been dated to the period when these tracks were first used; some eight to nine thousand years ago.[50] Surrounding this point is an area of some 153 square kilometres, centred on Stonehenge, in which there are more than three hundred Scheduled Monuments of National Importance.[51] The largest settlement of its day in Northern Europe was built at Durrington; and this was followed by Stonehenge. Something very special was happening here, and it was happening long before the stones arrived.

Artistic impression of a Neolithic house

In the later Roman period, all roads would lead to Rome. But before the Romans came, all roads appear to have led to Stonehenge. Wherever you came from, getting to Stonehenge was simply a case of finding the track that everyone else was on.

Stonehenge: The word

It is thought that the word 'Stan' comes from the old English word for stone.[61] This is the simplest and most obvious suggestion: many North European languages have a word for stone which sounds like the English word (examples are 'sten' and 'stein').

However, in Celtic languages from the Atlantic fringe, the word meaning 'stone' is quite different.[62] In Cornish, the word 'sten' means the metal tin. In Irish, 'stáin' also means tin and similar words for tin can be found in Catalan, Spanish and Latin. The word 'tin', with or without the 's', seems to be universal throughout Europe. Although the Roman Empire did not expand into all of these countries, this word may be derived from the Latin word 'Stannum'.

The word 'henge' is nowadays used in English to describe a circular structure, often a bank, ditch and/or mound sometimes accompanied by standing stones.[63] It was first used in 1932 by a British museum keeper, Thomas Kendrick, to describe a circular "sacred place".[64] Parts of all of these types of structure are found at Stonehenge, which ended up giving the name 'henge' to all the other monuments. Nobody knew what 'henge' meant and it seemed as good a word as any. As time went by, the word 'henge' has been more strictly defined and, bizarrely, Stonehenge is no longer classified as a 'henge'.[65]

William Stukeley, an early historian of the monument, thought that the word henge was *"plainly Saxon & signifys only the hanging stones"*: Perhaps a form of stone gallows.[66] But though hangings were carried out at Stonehenge in the middle ages, there is no evidence that it was built to be a place of execution.

On the other hand, a 'hinge' is something that allows a door to swing open. 'Hænge', 'Hengsel', 'Hing', 'Hinge', 'Eŋġe' and 'Gångjärn' are all words which mean hinge in North European languages. Similar sounding words for a 'hinge' exist in South European languages; for example 'Engoznar', 'Ganghero' & 'Engonçar'.

'Angel' can also mean hinge, pole or rod in other languages. Angleterre is the French word for England; Land of the Angles (a people of Saxon origin). Whatever language one uses in Europe, the word used to describe England is likely to sound like the word for hinge in that language.

2: AN INTRODUCTION TO STONEHENGE

Some European words with similar sounds:

	Hinge	*Rod (pole)*	*Angle*	*Tin*	*Stone*
Swedish:	gångjärn	stång	vinkel	tenn	sten
Irish	hinge	slat	uillinn	stáin	cloch
Cornish	---	---	---	sten	---
Scottish Gaelic	---	---	---	staoin	cloiche
Norwegian	hengsel	stang	vinkel	tinn	stein
Breton	---	---	angle	staen	(men)
Dutch	hengsel	stang	hengelen	tin	steen
German	angel	stange or angel	winkel	tin	stein
Faroese	hongsl	stong	vinkel	tin	steinur
Latvian	eņģe	stienis	leņķis	---	---
Danish	hængsel	stang	vinkel	tin	sten
Estonian	hing	---	vinkel	tina	---
Romanian	---	---	---	staniu	---
Latin	---	---	---	stannum	lapis
Welsh	---	---	ongl	tun	---

--- word not found or word in that language not similar

Curiously, all these words seem to be related; almost as if the words Stone and Henge are deeply embedded in European languages; and perhaps meaning something to do with hinges, poles, angles, tin and stone.

A history of discovery

Stonehenge was first mentioned in a deed of 937 AD,[71] but very few have been certain what purpose the stones served. The diarist Samuel Pepys wrote *"God knows what their use was"*,[72] whilst Daniel Defoe defeatedly recorded that *"all that can be learn'd from them is, that here they are"*.[73]

Though no Roman historian mentions the monument,[74] in the Historia Anglorum of 1129-30 AD, Henry of Huntingdon wrote of *'Stanenges'*, the *'second wonder of Britain'*,[75] that *"no-one can conceive how such great stones have been raised aloft, or why they were built here"*.[76]

In about 1620, King James the First requested the architect Indigo Jones to record the stones and find out their mysteries. Jones became convinced that it was a Roman temple and that one extra trilithon had once existed:

Illustration by Indigo Jones[77]

At the instigation of King Charles the Second, John Aubrey started another survey, largely due to a Court argument about whether it was a Roman temple or a coronation place of Danish kings.[78] Aubrey discovered both that Indigo Jones' survey was not very accurate and that an outer ring of mysterious holes had once existed:[79] Fifty-six of these holes are located at the perimeter and are known as the 'Aubrey holes'.

Following Aubrey, William Stukeley, a friend of both Edmond Halley and Issac Newton, re-surveyed the stones and realised that they were aligned to the Sun's position along the horizon at solstice. Stukeley was the first to suggest that Stonehenge dated prior to the birth of Christ and that it was a Druidic Temple.[80] He also discovered two cursuses near Stonehenge: wide and very long areas, with banks of chalk either side, which were later found to have been built much earlier than the stones.

2: AN INTRODUCTION TO STONEHENGE

John Wood, an architect, also surveyed the monument at about the same time and vehemently disagreed with Stukeley; thinking it to be a 'Druidic University' filled with Greek scholars.[81]

In 1877, Flinders Petrie allocated a number sequence to the stones, which is still in use today:[82] His system was to start at the north-east entrance and to number the stones clockwise. Charles Darwin, who visited the site in 1888,[83] also pitched in and suggested that earthworms had played a major part in the submergence of some of the stones. By now, the gentry of England had acquired a fascination for Stonehenge.

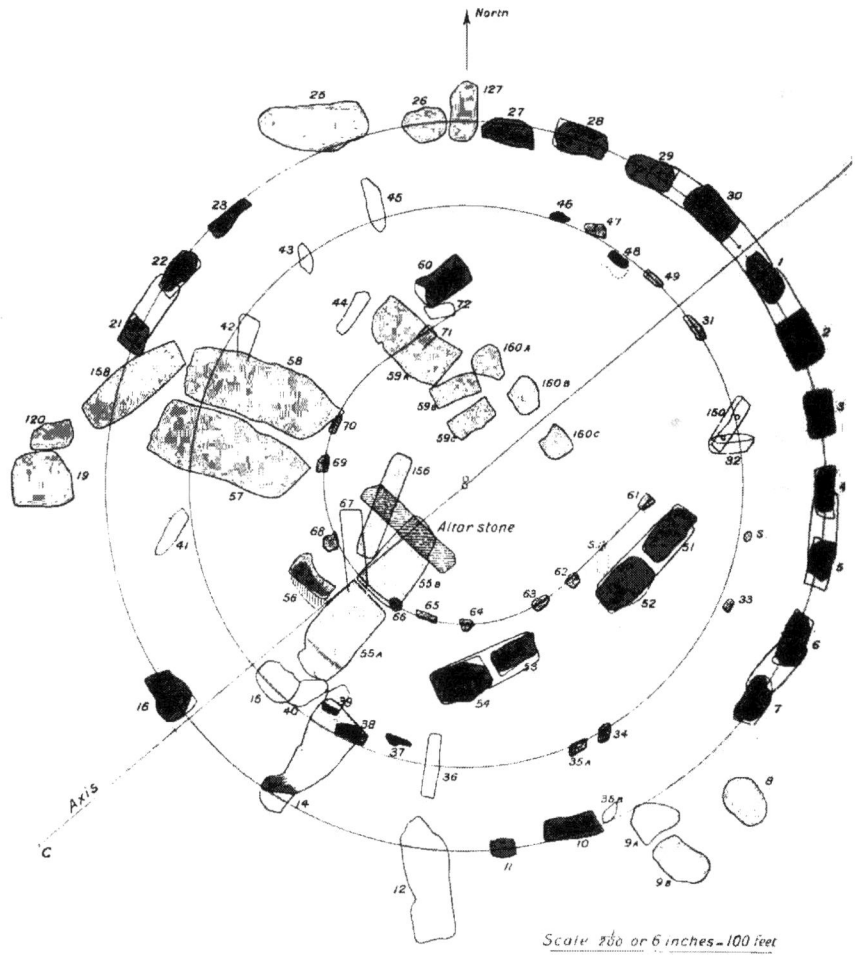

Barclay's plan of 1895 (Plan I) [84]

In 1900, one of the sarsen stones of the outer circle, together with its lintel, fell over. With growing concern, plans for restoration works were immediately put into place. Professor Gowland's straightening of the Great Trilithon[85] was undertaken the next year followed by several other restorations in the following decades. In Gowland's opinion, Stonehenge was *"a place of sanctity dedicated to the observation or adoration of the sun"*.[86]

Barclay's vision of Stonehenge when restored (1895) [87]

At the end of the Great War, in 1919, a certain Colonel Hawley was charged with excavating an extensive area of the site. It would not be unfair to say that modern archaeologists are not very impressed with the colonel's methods. A lot of information appears to have been mislaid and his finds are not always well recorded. Half a decade was to pass before the archaeologist Richard Atkinson became the last person to excavate large parts of the site. Even today, only about half of Stonehenge has ever been excavated.[88]

In 1995, the excellent volume 'Stonehenge in its Landscape' was published, bringing together all the information available at that time. However, this is a large technical volume and not for the faint-hearted. More recently, research and excavation in the neighbourhood of Stonehenge, together with some non-destructive work at Stonehenge using new scanning technology, has been undertaken.

But the one thing that pushed Stonehenge even further into the limelight was Gerald Hawkins' suggestion that the fifty-six Aubrey holes had been used both to predict movement of the Moon and to predict eclipses of the sun.[89] Astronomers such as Fred Hoyle and John North followed later with even more detailed examinations.

Hoyle wrote: *"Nobody, as far as I am aware, has argued that the axis of Stonehenge is only accidentally coincident with the direction of midsummer sunrise".*[90] As proof, Hoyle noted that Sarmizegetusa, a superficially similar prehistoric monument near the village of Grădiste in Romania, is also aligned to a solstice horizon event, though the opposite time of day to the one at Stonehenge.[91]

However, whilst Sarmizegetusa may be aligned to solstice, no solstitial alignments have been found locally at any of the other great 'henges' of Wessex.[92] More recently still, Professor Clive Ruggles reviewed all these claims and concluded that only a few of the 'Stonehenge alignments' can be explained satisfactorily. Clive's view seems to be that Stonehenge was not an observatory.[93] In the last few years, Anthony Johnson reviewed Stonehenge's symmetry and concluded: *"My own opinion is that Stonehenge has only one alignment, i.e. the major line of symmetry established along the line of the summer-winter solstices".*[94]

From the excavations within the Avenue, it is thought that stones may have existed long before Stonehenge was erected. And some form of early 'solstice alignment' is now thought to have been in place centuries before the stones of Stonehenge were put up.

So a puzzle that remains is whether or not Stonehenge, the place that we know now, was purposefully aligned to solstice. If it was, what purpose could have been served by building such a large monument in a position that blocks the view in both directions?

3: A POTTED HISTORY AND THE THEORIES

In about 2500 BC, a massive forty two-acre town was being built close to Stonehenge. This was to become the largest settlement of its day in Northern Europe.[01] It is thought that the people of this town, now known as Durrington, also built Stonehenge.[02]

About five thousand years before, a tree and three high posts, in a line some 37m (120') long, were put up some 250m (820') north-west of Stonehenge. This can be found within what used to be the old car park,[03] and marks the beginning of whatever Stonehenge stood for. These posts were up to 70cm (≈2¼') in diameter[04] and made of pine; probably selected from the local pine and hazel woodland of that early period.

These posts are now dated to some ten thousand years ago, with only one exception (post hole 'B') having a later carbon-date.[05] However, a tree can be several hundred years old and, if one sample is taken from the old part of a tree and another from the young part, carbon dating could show different dates.[06]

A potted history of the monument's phases

Professor Timothy Darvill uses the phrase *"Tamed Wildwood"* to describe the landscape of the following period (from about 4000 to 3000 BC.)[07] Pine had given way to a mixture of trees such as elm, ash, oak, hazel and yew[08] with large felled areas of open grassland and some localised farming. At about this time, the Lesser Cursus, and the Great Cursus, were built towards the north-west and the north of Stonehenge.[09]

This was a period in which cursuses had become fashionable[10] throughout the British Isles; though not on the continent.[11] The earliest cursuses are in Scotland and generally fall within the date range of 3660-3370 BC.[12] One curious feature of cursuses is that, typically, they do not *seem* to align to anything.

In the same period, leading up to 3000 BC, other monuments are known to have been exactly aligned to solstice. An example is Newgrange in Ireland.[13] From our knowledge of Newgrange, other alignments such as those of Knowth (towards equinox) have been deduced.[14] But these are rarely an exact match: a few early cursuses point in the general direction of a solstice event, but of the more than a hundred or so other cursuses and passage tombs in Britain, very few have obvious alignments.[15]

The timing of the construction of various stages at Stonehenge was recently revised[16] and shows a significant modification to what was previously thought to have been the construction sequence.

Dates are arrived at by using the decay rate of carbon-14 (C-14); an unstable, but relatively long-lived, hybrid of carbon-12 formed in the upper atmosphere. As the Sun's cosmic rays cause atoms to be broken, a small portion of the broken atomic parts join together, more or less at random, to form carbon-14.[17] The atoms of C-14 then join with others to form molecules which gradually disperse down into the lower atmosphere, where they are converted by plants using sunlight (a process known as photosynthesis).

At the time the plant dies, it stops taking in C-14; which then gradually decays into nitrogen. By working out how much C-14 is left in a plant, or in an animal which eats those plants, the age of the sample can be estimated. The way this is done is reasonably accurate and relies on knowing how much solar activity occurred in the past. However, the method is statistical and can be subject to quite a lot of error.

As the fourth millennium BC ended, a new era of continental weather conditions arrived: drier and warmer summers with colder winters.[18] By 3000 BC, the area around Stonehenge contained a lightly wooded and open grassy plain, not dissimilar to that which we can see today.[19]

At about the same time, something else was going on at Stonehenge: a large number of post-holes were being dug at its centre. Although much of the ground has not been excavated, over two hundred holes are known to exist.[20] The date that these holes were dug is not known, but a well known expert, Professor Stuart Piggott, thought there may have been a central timber structure on the site.

Prior to human-made constructions such as the post-holes, a series of natural geological features are also thought to have existed. Some of these appear to align with sunset at solstice.[21] Whatever happened to cause the solstice alignment to eventually be settled upon, it is possible that it was, as with the solstice alignment at Newgrange, a key feature of the early stages.

Newgrange: Ireland

In that era, some other similarities with the solstice-aligned Newgrange are evident. The Newgrange complex contains a major monument known as Knowth. This monument has two passages which both look towards the horizon. Both passages are slightly off-set from equinox; and one slightly more than the other. The passage nearest to the equinox 'alignment' is the longer of the two.

Before major construction took place, Stonehenge, like Newgrange, probably had an 'alignment' which pointed towards a solstice. Stonehenge's cursuses also both look towards the horizon. Like Knowth, both cursuses are slightly off-set from equinox. And one, the Lesser Cursus, is off-set slightly more than the other. In addition, the Greater Cursus, nearest to the equinox 'alignment', is the longer of the two.

Potted history: Stage 1 *(3000-2920 BC)*

The first major building phase included a circular embankment, which could have been up to two metres (6') high, and was built to surround the central area. The bank stands on a base 6m (20') wide and has a diameter of almost 100m (≈300').[31] Outside the bank there is a ditch with a small second bank of chalk beyond (known as the counter-scarp), which could have been up to 75cm (2½') high.[32] The ditch between the two is the easiest to date using carbon-14 because of antler pick-axes left behind by the workers.[33]

Below the bank there exist post-holes of some 20cm (8") diameter and 50cm (20") apart, possibly indicating an earlier phase.[34] There were two entrances through this bank: one main entry at the north-east, some 12m (40') wide,[35] and a smaller 3.5m (≈12') entry to the south.[36] This second entry is almost directly south of the centre. It is also possible that a third entrance existed to the south-west.

At this stage, existing geological stripes may have been visible along the 'solstice alignment'. It is also possible that one or more stones existed in a line; very slightly off-centre relative to a 'solstice line'. A series of timber posts may also have existed at this early stage,[37] close to the centre of the entrance. These posts were roughly in the position of the Moon's alignment at its furthest position on the horizon.

Artistic view of the early henge along the 'solstice alignment'

Within the inner bank, the Aubrey holes were dug[38] around about 3000 BC.[39] If these holes held posts, they may have been up to 10m (≈30') high. There were fifty-six of these holes; dug in a near-perfect circle and very accurately set out. These were cut some 0.56m (22") to 1.14m (45") down, with the deeper holes at slightly higher ground suggesting that the level of whatever went into those holes may have been important.[40]

The Aubrey holes vary between 0.74m (30") and 1.82m (72") diameter,[41] but show no alignment to the skies. As with the southern entrance, they are two or three degrees out from cardinal directions (north, south, east, west). Though the Aubreys may have been important, the amount of effort required to construct the circular embankment would have far exceeded the effort required to dig these holes.

The main entrance originally aligned several degrees north of solstice, but was at some point relocated so that it looked to the same general direction as the Avenue.[42] Along what later became that Avenue, one or more stones already existed, or were perhaps installed at a later intermediate date, in what appears to be the solstice alignment.

At this time, the cursuses and their banks would still have been the major feature of the area. These were, and still are, huge compared to Stonehenge's circular bank.

Artistic impression: The Great Cursus and Stonehenge seen to scale from above

Chalk plaques, with zigzag patterns similar in some ways to those found on stones at Newgrange and Knowth, have since been discovered in the area. These have been dated to a similar period to that in which the bank and holes were constructed (2900-2590 BC). [43]

3: A POTTED HISTORY AND THE THEORIES

Potted History: Stage 2 *(2620-2480 BC)*

The five huge stone trilithons in the centre of Stonehenge were most likely the next major element. This was probably followed by a double arc arrangement known as the Q and R holes. These holes probably consisted of a ring together with an internal arc at the north-east.[51]

26m (85') and 22.5m (74') diameter respectively, this ring and arc probably contained bluestones.[52] Each may have been topped with a lintel,[53] possibly to make an early version of Stonehenge (the bases of these holes have been found to contain dolerite chips).

If it was intended to eventually have two full rings, there would have been forty pairs of stones, together with a possible third partial mini-arc set inside these stones. It is not known whether or not these were abandoned before completion as excavation has only been carried out in the eastern and western quarters.[54]

At this stage, whatever was being constructed did not fully reflect either the alignment, or the symmetry, of what would come later. Other than the trilithons, if they were erected in this very early stage, this arrangement seems to have first pointed towards the centre of the old entrance. It also seems probable that the sarsen circle that we know today was erected some time after the trilithons.

Whilst likely, it is also not known whether the Q and R stones were immediately removed; this process may have occurred over a period of one or more hundred years.

The Heelstone, some D-shaped buildings and the Station Stone were also probably put in place during this phase. Beyond the north-east entrance, the Heelstone may have been moved from stone hole 97 (in the old line of stone holes) to its new position. As Mike Pitts recalls: *"we know (from my own excavation) that a megalith used to stand beside the Heelstone, and that after this was erected, and probably after it had been taken away, the Heelstone was surrounded by a small circular ditch"*.[55]

Stone hole 97 is substantial, 1.75m (5'9") across[56] and 1m (≈3') deep. Whatever existed there had been part of a row which roughly pointed towards solstice, but could also have pointed to a major moon stand-still.

For reasons that are not known, this alignment (along what would become the Avenue) was removed. The only major stone along this line (the Heelstone) was probably relocated to a new position in which it has remained to this day. As the archaeologist Aubrey Burl (quoting Neil: 1975) explains: *"It is always assumed that it was from the centre of the sarsen circle that the observations were made, but, from that position, as pointed out in 1975, 'the sun ought to rise to the right at the top of the heelstone on the morning of 21 June our time. But it does not; it rises about a foot and a half to the left'"*.[57]

The Heelstone's new position as seen from along the centre-line

Many people have suggested that the name of the Heelstone was derived from the Greek word for the Sun: 'Helios'. If it was, it is strange that the builders would choose to move the stone so that it did not align with the sunset at solstice. Interestingly, the English word wheel comes from the Old English word hwēol, which derives from the meaning 'to revolve, move around'. Similar words also exist in Northern European languages (Swedish, Norwegian, Dutch and Danish). For example in Norwegian, the word 'hjul' means 'wheel' and 'hæl' means 'heel'. As a verb, the word heel is a synonym of the verb angle (as in 'heeled the sail'). So this word is also closely related to the meaning of 'hinge', which itself repeats across Northern European languages.

Potted history: Stage 3 *(2480-2280 BC)*

In the third stage, the Avenue was built (or, perhaps improved), the ditch cleaned and mounds raised around two of the Station Stones to form their north and south barrows. The Slaughter-stone companions, (in stone holes D and E and probably erected in the previous phase), were now taken down to widen the north-east entrance. A large pit was also dug into the north side of the Great Trilithon.[58]

However, whilst the Avenue had bluestone and sarsen chippings beneath its banks;[59] indicating that it was built up in this later phase, it is not certain that a less grand form of the Avenue had not existed earlier.

The Avenue: Barclay's extract from Stukeley (1895) [60]

It was thought that the Avenue had once been constructed in two phases. However, no evidence of this appears to have been found in more recent investigations: [61]

> "The stretch of avenue in which we were most interested was where the magnetometer survey had shown a kink in the ditches, suggesting that a new stretch of avenue had possibly been tacked

on to the end of the 500-metre-long straight section. In reality, we could see on the ground that there was no such deviation: the bend joined seamlessly with the straight section."

In addition, the Avenue bend was not rutted, leading to the conclusion that it was unlikely to have been used for processions.[62]

Potted history: Stage 4 & 5 *(2280-2020 BC and 1680-1520 BC)*

The Q and R stones were then rearranged into the outer bluestone circle (between 2270 and 2020 BC) and a bluestone oval set within the trilithons (2210-1930 BC). The inner bluestone 'horse-shoe' may be all that remains of the oval and its stones, repositioned from the earlier Q and R ring.[63]

Nearly a thousand years after the stones were first erected, two circles of rectangular holes, known as the Y and Z circles, were constructed in about 1640-1520 BC.[64] Located just outside the sarsen outer ring, there is no evidence that these holes, apparently separated by a period of between one or more hundred years, ever held stones.[65]

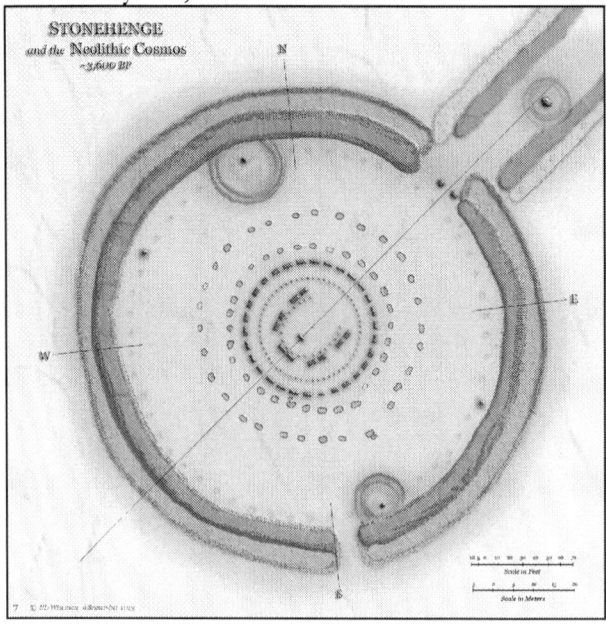

The Y and Z holes. Image © ND Wiseman 2013

The Stones: A potted summary

Stonehenge contains well over one thousand tons of stone.[71] The majority of this is sarsen; a fine-grained type of siliceous sandstone[72] which is formed under extreme ground pressure. Grains of quartz sand become bound together with siliceous cement[73] to form a stone which is harder than granite. In 1654, the diarist John Evelyn wrote of this material: *"The Stone is so exceedingly hard, that all my strength with a hammer could not break a fragment"*.[74]

The smaller bluestones, on the other hand, are a much softer material. The glaciers of the last ice age never got as far as the Salisbury Plain,[75] so the bluestones are generally thought to have been hauled from the Mynydd Preseli area of Wales [76] by gangs of workers.

Geoffrey of Monmouth, in his 'History of the Kings of Britain',[77] wrote that the Stonehenge Circle was originally erected in Ireland from stones brought from Africa. However, it is very much more likely that the larger sarsen stones were brought from the nearby Marlborough Plain. That area is some twenty miles to the north, near the larger (and earlier) sarsen monument of Avebury.

The outer perimeter of Avebury today

The sarsens were probably found as large stones buried below ground[78] and seem to have been specifically sought out by the Stonehenge builders. Given that Avebury already existed, and that there is lots of sarsen stone available in that area, this type of stone would have been well known. How these stones were transported, over almost twenty miles of uneven countryside,[79] remains a mystery.

Stonehenge is one of the only monuments in Britain where megaliths were carved. It is likely that these were made by flaking the natural blocks to get a rough shape[80] and then chipping away at the surface using mauls[81] weighing up to 26 kilograms.[82] The distribution of flakes shows that the bluestones were dressed inside the circle, whereas the sarsens were worked outside the ditch and bank.[83]

At the perimeter of the main stone circle, there once existed up to thirty of these massive uprights; not all equal in size but placed with their centres at almost exactly the same distance.[84] Typically weighing 25 tons, these uprights average some 4.1m (13½') high[85] and are topped with lintels weighing up to 6 tons.

There was an obvious intent to make the circle perfect. Some uprights are longer than others and were buried to the depth needed to make an exactly level top-most ring.[86] After the stones were erected, the shaped lintels were then raised into position.

The smoother faces of the sarsen circle uprights look inwards and are set on a true circle of just under 30m (≈100') in diameter.[87] First noted by Stukeley, these sarsens are more finely, and smoothly, worked on the inside[88] and, to a small extent, dressed to be gently tapering upwards (an architectural device known as entasis). This effect might have been used to help create the optical illusion of straightness.[89]

Stones of the outer circle drawn from slightly off-centre

Dovetailed and attached with mortice and tenon joints, the stones which were to become lintels are also finely worked on the inner faces and then placed to become the precisely level top surface.[90] Whilst not quite as good as we can achieve today, the accuracy to which this monument was set out is quite astonishing.

Within the middle of the monument, five sets of paired upright stones exist and, onto these, massive lintels were lifted. These sets of three stones are known as 'trilithons' and are arranged symmetrically about the single largest set of three; known as the Great Trilithon. The largest of the uprights weighs in excess of 40 tons, and each was carefully worked, with considerable effort, to make one face as smooth as possible. With only one exception, these smooth faces look inwards.[91] The Great Trilithon is unique in having its smoother, flatter face look out rather than in.[92]

Atop of the trilithon uprights, the lintels are about 1m (≈40") deep and 1.2m (≈47") wide; except for the Great Trilithon lintel which is only 70cm (27") deep. The tops of those lintels are 6m (≈20'), 6.25m (≈21') and 7.5m (≈25') above ground[93] and, in all cases, the lintels have been shaped and then fitted onto the uprights with mortise and tenon joints formed in the stone.[94]

A lintel of one of the trilithons

The Great Trilithon was tilted back to position at the start of the last century, with the result that its new position is slightly rotated; and also further away from the centre of the monument by as much as 60cm[95] (≈24") from its original location.[96] A huge pit was later extended from the Great Trilithon and, allowing for this, these trilithons have been dated to 2620-2480 BC.[97]

The sarsen stones of the monument are largely symmetrical with but two curious exceptions: Firstly, a smaller than usual stone exists in the outer circle. This is directly south of centre, perhaps making a larger than usual opening. Secondly, one Trilithon pair stands out as being unique:

Stone 54 stands directly south of the centre and has a 'foot' about twice the width of the rest of the stone.[98] Despite being shorter than the Great Trilithon, it is the heaviest stone. This upright was also constructed in a unique fashion: it was packed with a hard compact rock which *"...cannot, as yet, be referred to any of the existing stones, was used as packing for stone 54".*[99]

Set just inside the outer sarsen circle, a smaller ring of stones known as the 'bluestone circle' are arranged with a diameter of 23-24m (75'-79').[100] These stones are variable in size and shape; some 2m (6'6") or higher above ground[101] with a spacing of about 2.7m (9').[102]

Most of the stones in this circle show little sign of having been dressed.[103] However, Stone 36 was found partially buried and this stone does show indications of having been a lintel: it was beautifully worked and the mortice holes are too close to centre to be a lintel ring.[104] If it was a lintel, this stone would have been part of an early 'trilithon configuration'.

Within the sarsen horse-shoe, another set of smaller blue stones, up to 2.5m (8') high and spaced 3.5m (11½') apart,[105] were arranged as an oval and graded in height. The shortest stones occur at the horse-shoe edge and the highest are towards the back (near the Great Trilithon).[106]

Originally, this was almost certainly an oval shape, which was modified at later date.[107] It contained the most finely worked of the bluestones.[108] At least two abandoned arrangements, from the remains of earlier oval patterns, were found in front of the horse-shoe.[109]

Bluestones within the inner monument

The axis of the monument

Where architects can, they will design buildings to be symmetrical: symmetry is more pleasing to the eye. Even a standard English semi-detached house will almost always be symmetrical about the division wall between the two properties.

Stonehenge too is a largely symmetrical monument and, from its symmetry, the axis has been established. However, there are a few exceptions to the symmetry of this stone monument. Perhaps clues to its meaning can be found from looking at what does *not* fit this symmetrical pattern.

Long before the stones were erected, the main opening through the great henge's bank wall pointed more towards the north. A series of post-holes, probably holding timbers, existed at and outside the entry[111] (anti-clockwise of the eventual stone alignment). At some point, the bearing was shifted to 49.9° from north (originally 46.55°)[112] and a series of stone holes known as C, B and 97 probably came to signify the choice of the final symmetry of the monument's axis.[113] At this stage, stone hole 97 may have held the Heelstone (which is now located in stone hole 96.)[114]

The Slaughter-stone, which fell over in centuries past, stood upright[115] some 5m (16') above the ground. This stone is estimated to weigh 28 tons and is located close to the entry of the 100m (330') diameter henge.[116] Unlike many of the other out-lying stones, the Slaughter-stone was made to shape.[117] Two further holes, known as D and E, are some 4m (13') apart and, together with the Slaughter-stone, may once have symmetrically framed the entry.[118]

The fallen Slaughter-stone

However, stone hole 96, the eventual location of the Heelstone, is an exception to the rule of symmetry. It cannot be established that 97 and the Heelstone (96) were a pair[119] and, in fact, it looks very unlikely that they were. A ditch,[120] which cuts through the hole for 97, was later dug around the Heelstone, showing that the two holes were not of the same era. Stone hole 97 is 5m (≈16') long and might once have held the Heelstone, which would then have been moved to its current position and only later had the ditch constructed around it.[121]

A heel is the point of the foot about which you rotate. It gives rise to the English phrase 'turn on one's heel'; meaning to make a sudden change of direction. Over four thousand years ago, something curious happened at the location of the Heelstone: its alignment was changed, and specifically modified, so that it no longer aligned with everything else at Stonehenge.

The four Station Stones, located in the same boundary circle formed by the earlier Aubrey holes, are arranged as an almost exact parallelogram about 80m (262') by 33.5m (110') wide.[122] Only stones 91 and 93 survive, but the two missing stones (92 and 94) had low chalk mounds around them; together with encircling ditches. These were originally thought to be for burial, so are still known as the north and south barrows.[123]

These Station Stones, or the remains of their holes, are said to be approximately aligned on the most southerly moon-rise and most northerly moon-set.[124] However, at this latitude, any box, or any rectangle drawn on the ground, will align to these lunar events if one edge of the box points towards a solstice sunrise or sunset).

A further suggestion was that the diagonal between stones had some sort of astronomical significance; thus explaining why they were not arranged in the shape of a square or a line. However, the astronomer Clive Ruggles notes:

"The question of why the station stones were placed in a rectangle and not a square has given rise in the past to the tentative suggestion that the WNW-ESE diagonal might have had astronomical significance; however its declination -16°, has no obvious explanation in terms of the Sun or Moon and in any case the diagonal passes across the centre of the site where it might have been partially obscured by the bluestones, and certainly would have been obscured by the later sarsens". [125]

Theories

"The starting point is often not the body of archaeological evidence at all, but a theory which forms a mould into which all the disparate elements appear to fit". (Anthony Johnson.) [131]

As with the nearby monument of Avebury,[132] some researchers believe that Stonehenge was associated with death. Early theories include that it was a work of the Phonecians, a temple of the Druids, a monument of the British to Anaraith (a goddess), a monument to Queen Boadicea, a Roman temple, the burial place of Uther Pendragon and other kings or a Danish monument.[133] No doubt there were many others.

In more recent times, and noting the Druids' belief in reincarnation and their known fascination with astronomy and the size of the world, Hancock[134] suggested the possibility that Stonehenge may have been connected to a spiritual quest for reincarnation and immortality of the soul. Terence Meaden suggested that Stonehenge was seen not just as a temple, but as a goddess in itself.[135] All the early, together with some of the more recent, theories are of a 'ritual' nature.

Gerald Hawkins famously proposed that Stonehenge was a type of observatory of the stars.[136] However, Professor Hoyle, who advocated a similar position, remarked that *"When I first read Professor Hawkins' book Stonehenge Decoded, I was struck by the angular differences between the actual measured sight lines and the astronomical alignments"..* (as shown in Prof Hawkins' book). *"Even with primitive equipment, a stone could still be placed to within about a foot".*[137] He also said that *"In one case in six, a purely arbitrary direction would happen by chance to agree with an astronomical alignment".* [138]

More strikingly, Hoyle notes that no account of the achievement has been made in *"the full light of documented history".*[139] This is perhaps key to the mystery: if something spectacular had happened at Stonehenge, there would be some sort of record, even if only in mythology.

Though the astronomical observatory theory is the most popular and well known, Professor Clive Ruggles has recently concluded that there is little, if any, evidence of intentional astronomical orientation in the early work, [140] nor evidence that the structures at Stonehenge *"deliberately incorporated a great many astronomical alignments, or that they served as any sort of computing device to predict eclipses."* [141]

The observatory theory seems to be dead as a result of Ruggles' work. This is not really a surprise given that Stonehenge's stones face inwards; not a direction that could have been any help with the observation of objects that are outwards. Nevertheless, the earliest version of this monument did seem to point towards a 'solstice alignment'.

No theory has yet come to explain what Stonehenge was for. Some of the biggest mysteries are:

- What is the internally facing structure for?
- Why is it arranged with outlier stones and a huge bank?
- Why is there no record of what was done there?

In 2008, Anthony Johnson said: *"The starting point is often not the body of archaeological evidence at all, but a theory which forms a mould into which all the disparate elements appear to fit"*.

All of the features described in the chapter you have just read appear to fit the hypotheses described in the pages which follow.

4: MIRROR, MIRROR

Between 6000 and 7000 BC, copper was being hammered at Çatal Hüyük in Turkey.[01] The earliest mirrors, made of polished obsidian, have also been found at Çatal Hüyük and dated to about 6200 BC.[02] The earliest copper mirrors found so far date to about 4000 BC.

Extraction

Tin is a soft metal with a low boiling point. Experiments on streamed ore, undertaken during the 1970s by Friede and Steel,[03] showed that it can be extracted from cassiterite using simple clay furnaces and charcoal.

Another of Fried and Steel's conclusions was that early tin making processes would not produce a great deal of tin relative to the amount of tin available in the pebbles. However, the tin that was produced would be of a very high quality.

Tin is found as a heavy ore. The process used to refine it, known as smelting, is simpler than that used for copper, although the two processes are similar in many aspects. Fuel can be burned in a kiln to produce both heat and carbon monoxide. This gas reacts with the metal oxide to give off carbon dioxide, leaving behind a mostly pure and unoxidised metal. In essence, this is a fast reversal of the 'rusting' process.

Of the two early metals used to make bronze (copper and tin), tin is by far the easier.[04] That those two metals can be produced relatively easily makes then likely candidates to have been first manufactured from ore during the Neolithic.[05] However, tin is a soft metal whose use is limited for making tools. Copper, a slightly harder metal, does lend itself to making some types of tools. But bronze (copper mixed with a small amount of tin) is very hard and suitable for both tools and weapons.

The source of metal

Tin can be found as surface deposits in alluvium where the ore (cassiterite) has deposited within shallows set within the landscape. As granite at high level is eroded, some of the deposits of ore occur as small pebbles (shodes), which are rich in tin.

An easily accessed, and ancient,[06] tin ground can be seen at Haytor in the Dartmoor National park:

Tin streaming ground at Haytor, Dartmoor

Below its tors, a number of streams join to form the River Lemon:

Streams leading to the River Lemon, Dartmoor

Though the exact method used is unknown, the early mining of tin, using those streams, would have been a relatively simple process. The stream is diverted sideways and "streamed" over the adjacent ground, probably using a raised timber channel. This process allows the soils to be dug, and alluvium washed away, along the newly created watercourse. That method leaves behind rocks, ordinary pebbles and the heavy shodes.

Dry banks and trenches at tin streaming ground, Dartmoor

With the shodes collected, the waste rocks can be collected and dumped next to the diverted stream-trench and the next level down 'streamed'; to as deep as the miners dared to go. The stream is then diverted just above the line of the original 'streamed' diversion and the process re-started to form a new parallel trench. The remains of this process can be seen as dry banks and ditches, now showing themselves as lines within the bracken.

If the ground is particularly rich in tin ore, but that ore deep within the alluvium, timbers could be used to brace the banks along the line of the dig. This process needs deep trenches to carry the water away and results in the type of scarred landscape seen at Haytor; the remains of deep banks and trenches.

In the book 'Tin in Antiquity', Roger Penhallurick describes[07] how a large piece of timber, with an existing notch formed by hand tools, was found in 1839 during the mining of a nine-metre deep bed of tin ore (cassiterite) in Cornwall. Too young to have been deposited by nature, carbon dating showed that it was as old, or much older, than Stonehenge. Below is the full extract of that description, produced for The Institute of Metals:

> "Sub-fossil wood resting on top of the tin ground ought to have a date of c. 10000 bp, or even earlier. Wood found within the tin ground cannot have arrived there at the same time as the cassiterite, and it can hardly be doubted that Winn's oak trunk had been put there by man. It has been dated at the University of California, Riverside (UCR 1828), to 4140 +/- 100 bp, giving a time range of 3015-2415 BC. This is uncomfortably early, even if the precise calendar date lies at the end of the range. It suggests that tin streaming in Cornwall began much earlier than hitherto suspected."

This piece of oak was discovered at the Wheal Virgin tin streamworks, then at Levalsa Moor (located immediately below St Austell).

In summary, tin ore was almost certainly being mined long before Stonehenge was constructed. Because of the depth of the oak found at Levalsa Moor, this was probably being done on a very large scale.

Polishing and the reflectance of tin

The spectral reflectance (the amount of light reflected back) of polished tin is typically in the region of 80% for unpolarised light,[08] reducing to some 70% or so if an oxide film is allowed to develop.[09]

From tests on castings, using natural materials easily available, I found that tin can be polished with ease using clay. However, the local clay in my area (the Weald) leaves small scratches. Crushed and ground chalk was found to produce a much better finish, and I also found that apple juice removes the golden tarnish which forms on tin over time.

A better alternative, Pipe clay, was at one time "the standard polish for naval and military tunic buttons" and "chalk, mainly from England and France and known as 'whiting' is used for hand polishing metals".[10] Of the two, Pipe clay would probably give a better finish than chalk.

Close to Stonehenge, there is "an irregularly-shaped ditched enclosure.."[11] which "sits among traces of a later prehistoric field system". This can be found adjacent to Clay Pit Hill in the area of

Chitterne. According to local historians,[12] this pit contained "smooth white clay with round pebbles [which] had been uniquely preserved in the chalk plateau by sinking into a large hole, of unknown depth, in the chalk millions of years ago. The nearest equivalent clay is to be found 14 miles to the south-west" ... "The clay is thought to have been known and used by man at least since neolithic times. Two flint scrapers, thought to be neolithic, have been found at the site."

From the same source,[12] the Rev. John Thomas Canner, vicar of *Chitterne All Saints with St Mary* from 1904 until 1925, is reported as saying that "the hill called Clay Pits is the place where the best clay in England is to be found for the manufacturing of tobacco pipes."

From the local history above, the Stonehenge area appears to be one of the best locations to find the natural materials (chalk and Pipe clay) that would be needed to polish and finish mirrors made of tin.

Antiquity and Stonehenge

Professor Gowland, whilst re-erecting Stone 56 (part of the Great Trilithon), found traces of copper carbonate some seven feet ($\approx$2m) down.[13] This relic of a long lost object indicates that copper was in use during the time that the foundations for the largest sarsen stones were being dug.

In addition, two chalk blocks discovered at Durrington have long thin V cuts showing the possible use of a metal axe. The ditch in which these were found dates to 2480-2460 BC. That 'Beaker' copper composition also hints at a date one or two centuries earlier than 2400BC. In addition, the change in the size of trees felled around Stonehenge shows that metal technology had been introduced prior to its construction.[14] By 2500 BC, metal technology, found in and around the Stonehenge barrows, also shows that extensive trading networks had become established within the region.[15]

In 1586, a tin plaque was found at Stonehenge and noted by the traveller William Camden: *"In the time of King Henrie the Eighth, there was found neere this place [Stonehenge] a table of mettall, as it had been tin and lead commixt, inscribed with many letters, but in so strange a character that neither Sir Thomas Elliot, nor master Lilye (School master of Pauls), could read it, and therefore neglected it."* The writer Stukeley was later to say: *"But eternally to be lamented is the loss of that tablet of tin, which was found at this place, in the time of King Henry VIII (the Æra of restitution of learning and of pure religion) inscirb'd with many letters but in so strange a character... No doubt it was a memorial of the founders, wrote by the Druids, and had it been preserv'd till now, would have been an invaluable curiosity."* [16] At the time, the word 'tin' was common parlance for pewter (tin and lead), and this may explain why the tablet had survived.

Further circumstantial evidence, from archaeological investigations,[17] indicate that metals were in use at Stonehenge during the time that it was constructed.

In the book; 'The Early British Tin Industry', its author Gerrard notes that: *"Detailed electron probe micro-analysis of European Bronzes has led Professor Northover to suggest that "the number of metal sources used at one time was very limited, and there was often only one". Consequently, trading of metals was on a large scale and this phenomenon he termed the 'metal circulation zone'. For the Early Bronze Age, Northover has noted that tin bronzes were probably exclusively produced in Britain from South Western cassiterite."* [18]

In other words, all early tin throughout Europe probably came from one place. That one place was probably Cornwall.

More recently, Berger et all[19] analysed early tin from five sites in the eastern Mediterranean area. That analysis of the tin isotope composition helps to narrow down the tin's origin and, in combination with the trace elements found, points to Cornish tin ores (possibly from the "Carnmenellis granite area") as the most likely source.

The Carnmenellis granite area is at the far end of Cornwall. The next large outcrop of granite is at, and above, St Austell (known as the St Austell granite complex). Levalsa Moor, the place that the 4,500 to 5,000 year old pre-notched timber was found; buried deep in the ground, is located at the boundary of this area.

The evidence that metals were in use, long before Stonehenge was built, is therefore quite extensive. Copper and tin are known to have come from Cornwall, only 200-300km to the west, and, as a result of recent findings, the "Copper Age" is now thought to date back to 2500-2200 BC.[20]

5: STONEHENGE AND THE HINGE

The images in the following two chapters were produced using a three dimensional computer model which was based on a newly invented renewable energy device. The layout, and features, of the stones at Stonehenge were found to replicate the model's design requirements.

Stonehenge was built at the dawning of a new age. In Egypt, the Pharaohs would soon start to build pyramids and, in Britain, metals technology had just been introduced. A few hundred years later, tin and copper would be mixed to form bronze. With the discovery of alloys, the British Bronze Age would start and the Stone Age would become a thing of the past.

It has recently been discovered that metals were in use at the time Stonehenge was built. Metal has unique properties which, in addition to making good weapons, can be used in inventions. One such invention, a hinged mechanism which concentrates light, fits precisely into Stonehenge's structure. This light-concentrating system, described in the chapter following this one, could also be used to demonstrate how the Sun seems to move around the Earth; if it were believed that the Earth was fixed at the centre of the Universe.

But to begin, Stonehenge's overall plan layout is shown to be the same as an idealised geocentric description of the Universe. In this chapter (chapter 5), every feature of the plan layout of Stonehenge will be shown to be explainable using an old 'geocentric' way of thinking that was only abandoned by science in the 17th century. The introduction will start by describing how the search for knowledge could have resulted in an early fundamental view of the Universe; and the subsequent creation of Stonehenge itself.

The times

At the time Stonehenge was built, it is believed that life revolved around herding, farming, hunting and gathering; with some communities also producing pottery, wood-crafts and other skilled work such as the management of woodland for fuel and building. Mining, smelting and metalwork were newly discovered technologies.

Although science was in its infancy, the people of the time may have felt that they were living through the first industrial revolution.

Cornwall was a major source for tin ore (Cassiterite) and mines elsewhere could have produced copper and perhaps lead. But these new materials would have come at a very high cost. To make metal from ore, charcoal would have been made from a dried coppiced timber such as hazel. The ore has to be mined, smelted and then probably refined, once again, using a re-melting process. After smelting, metals are often re-worked using heat.

All these processes are relatively dangerous work. At this time, metal objects would probably have taken more effort to produce than anything made of the old materials such as timber and stone. Shaping a large rock using a stone maul would be relatively easy compared to the effort and expertise required to make one saw. Whilst those new technologies were being developed, storage of food would have remained critical to survival because prolonged winters could signal starvation for the community.

Long after Stonehenge, the Romans and Greeks believed that the world was at the centre of the Universe. Only recently have we come to accept that our world travels around the Sun. But on a world believed to be located at the centre of everything, the Sun does not seem fixed at all; instead, it appears to orbit the North Pole in summer before gradually moving to the south, where it spends the winter.

In the modern era, a discovery mission has sent the rover, 'Curiosity', to the planet Mars. Similarly, peoples of the past must have considered the whimsies of the Sun and Moon worthy of their curiosity. These objects seem to move within the heavenly firmament, yet are seemingly not committed to fixed positions within it. If the Sun were thought capable of making slight changes to its own yearly cycle, the perception may have been that starvation could result from the Sun's action. If this were believed possible, inventions that could help to understand the heavens would be borne out of necessity, not curiosity.

Astronomy on a fixed world

Our world appears to be solid and fixed. If you were to be placed in a new computer-generated Universe, somewhere which looks exactly like Southern England, there would be no obvious way to tell if you are on a disc, a ball, a cylinder or a flat endless plain. The Universe beyond our world might also be unknown; it could be a solid sphere, it could be stars with space between, or it could be something else entirely.

Simple experiments can help to show the nature of the world and the heavens. For example, the stars above us can be seen to move over the course of a night. A diligent observer, or someone with far too much time on their hands, will soon notice that some stars seem to move less than others.

By taking straight sticks, and pointing them at those stars which appear to move the least, the polar axis of the world can be found. On returning a few hours later, that one stick which still points to the same unmoving star (or dark spot in some ages) marks the polar axis around which the rest of the stars rotate.

The experiment above finds the fixed point of the heavens. The stick which was used to find it points in the direction of what we now call the polar axis. At the time Stonehenge was built, this point was marked by a star called Thuban. But prior to Thuban, there would have been no obvious marker for thousands of years.

Using the polar stick, another stick can be tied with string to make a sail which can be rotated like a hinge:

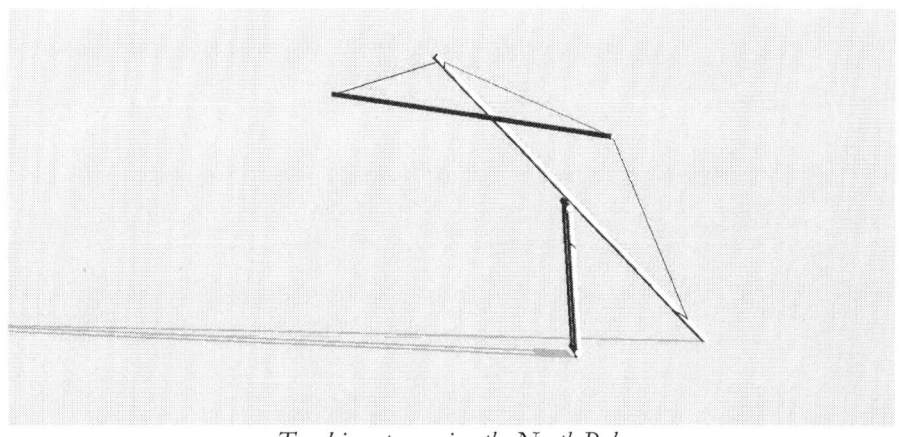

Tracking stars using the North Pole

Looking from the ground, this sail can be rotated to line up with other stars. As time goes by over the course of a summer night, the sail can be rotated slowly to keep pace with one star. Other stars will then remain at the same position along the length of the sail. By counting, and moving the sail in small angular steps, it is possible to show that the skies rotate like a 24 hour clock:

The clock of the stars

The experiment above shows that the Universe revolves. It does not show what shape the Universe has; a revolving ball would look much the same as a revolving cylinder; which also looks the same as stars with space between them.

By looking down the length of the sail, and rotating it constantly and slowly (rather than looking from the base of the polar stick), other stars can also be seen to be keeping pace. If the sail is at right angles to the pole, these stars will be near to the equatorial axis and are unique because they rise to the east, and set to the west.

The discovery of east and west stars can be very useful. By noting where equatorial stars set, perhaps from a location on high ground, the direction of something else (for instance a known destination such as a village) can be found:

5: STONEHENGE AND THE HINGE

Following the stars

In the Northern Hemisphere, the best place to do this experiment is on a north-facing slope. If the experiment is done in England, over a long winter night, the stars can be seen to rotate in almost a full circle. The most obvious explanation for this, if the Earth is fixed, is that the Universe would also be a sphere.

Even today, the most likely explanation for the Universe's shape is that it forms some sort of spheroid.

The sail of the stick can also be pointed towards the Sun. Using exactly the same experiment, the Sun can be shown to rotate around the same polar axis as the stars. In Southern England, at about 51° latitude, the Sun can be seen to rotate around a stick which points up to the North Pole at about 51° from the ground. (Latitude, the number of degrees from the equator, can be found anywhere in the Northern Hemisphere by pointing a stick at the North Pole and then measuring the angle down to flat ground).

During the spring and autumn equinoxes, a sail pointing to the Sun will be at right angles (perpendicular) to the pole-stick. To make it follow the Sun in high summer, the stick must point up from perpendicular by about 24°. At winter equinox, around about the 21st of December, it must point down by about 24°. Like the stars, if the rotating stick is made to follow the Sun, it also turns with the heavens; just like a 24-hour clock.

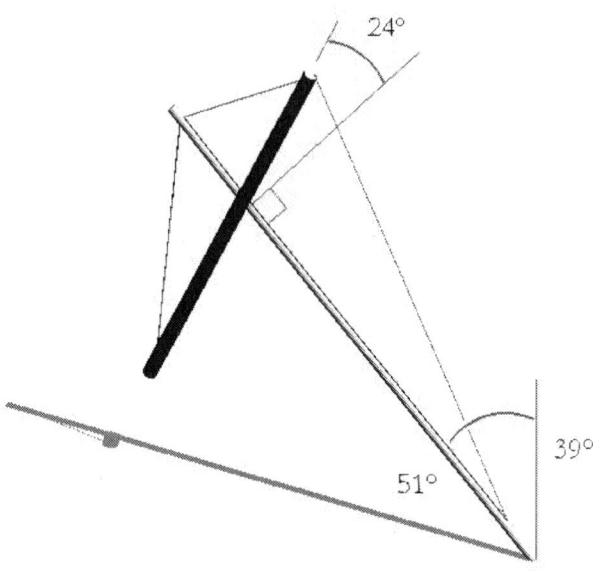

Summer arrangement

These sticks can show that the Sun moves just like the stars, with but one major exception: the Sun appears to gradually change its position in the sphere. In high summer it is 24° up from the equatorial stars and, in the depths of winter, it has moved to 24° down.

The same experiment can be done further north. The further north one goes, the higher the angle between the ground and the stick becomes. To quickly verify that this is not caused by the ground's slope, the stick can be held by its end and gravity allowed to make it point down (the angle formed on flat ground will be a right angle).

If the angle of the polar axis changes when going north, a logical explanation is that the shape of the world must also be changing.

If the skies are probably the shape of a ball, is the Earth a ball too? The changing angle of the pole star, when travelling north, seems to indicate that it is. Although it is not possible to see the curvature of the Earth at ground level, today we can see larger ships disappear over the horizon; also suggesting that the Earth is curved.

One way to test the idea, that the Earth is curved, is to find a high spot on an island far away from other land. The tip of a stick can be aligned with another so that both tips meet with the horizon. When looking back from the other stick, if both sticks still line up with the horizon the world must be flat and endless. However, if the second stick's tip is above the horizon, the world is probably either a disc, a cylinder or a ball:

The World if a disc (courtesy of NASA)

But there is a problem with this experiment: when it is tried on a hill looking over sea, the horizon seen during the day can be different from the one which appears when the Sun sets (or when it rises). This is due to haze, which can stop the real horizon from being seen during the daytime:

The effect of haze of horizons

A way around this problem is to find a location where there is coastline jutting out into the sea, together with a tall hill set within that promontory. At that hill, two sticks can be aligned over sea with the morning sun. At the place where the sea horizon is in front of the Sun, an exact line down to the horizon can be found. After perhaps weeks or months, the Sun will eventually set in exactly the opposite direction, allowing the difference between angles to be found.

In a location which looks over sea, directly east and west, the whole experiment can be done in one day. At equinox, the Sun rises in the east and sets in the west:

Aligning to sunrise

A high spot which works for this experiment can be found near Beachy Head at Bourne Hill (about 200 metres above sea level). There is a Neolithic mound with an almost flat top in exactly the right location:

Aligning to sunrise using tripods as sticks

This last experiment only shows that the angles to the east and west are the same. However, there are several ways to improve this (see chapter 7 onwards). The experiment above shows that the Earth is either a sphere-like object or that the experiment's location is, by chance, at the centre of either a curved earth-disc or, perhaps, some type of cylinder (though the change of angle to the polar axis, found when going north, shows that the world curves from north to south).

To find out which idea is correct, the experiment could also be done again at about the same height, but further along the coast. If the angles to the horizon start changing over these new sea views, the Earth could be a cylinder or a disc. If the angles never change (allowing for the height of each hill), the Earth is almost certainly a sphere. Other ways to double-check this idea also exist; and these appear to have monuments arranged in the correct location to make that happen (see chapter 7 onwards).

Other good locations in Southern England, for this experiment, can be found at St Catherine's (236m); the tallest hill on that part of the Isle of Wight which juts out into the English Channel. At its summit, there is an unusually large Neolithic mound which appears to be bowled or flattened on top (scheduled monument no 459799). [01]

By doing the experiment at the Isle of Wight, Bourne Hill and other places, the world on which we live can be shown to be curved, like a sphere, in all directions. Further along the coast, in Dorset, the experiment could be done at a 203m high point known as Swyre Head; where the coastline also juts out. An unusually large Neolithic bowl tumulus also exists at this high point (scheduled monument no 456525). [01]

More detail on these monuments can be found in Chapter 7 and onwards.

Swyre Head Tumulus

With a few simple experiments, the world can be shown to be round; though it can not be said for certain what the other side of the world looks like. The world below our feet appears to be solid and fixed, so if the assumption is made that the Earth is a fixed ball, the axis of rotation (of the moving heavens) can be shown on a drawing of that ball.

The Sun, appearing to change its orbit from winter to summer, can also be drawn onto that image of our Cosmos:

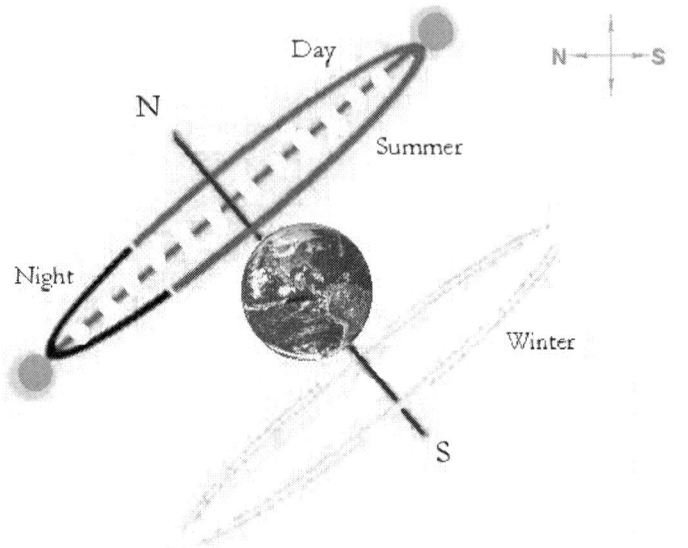

A drawing of apparent geocentric movement of the sun.

The most likely explanation for the heavens is that they are also a sphere, but as yet no proof is available to show how big the heavens are.

5: STONEHENGE AND THE HINGE

The Hinge of the Heavens

The hinge of the heavens, and the discovery that the world is a sphere, would probably be important enough to deserve to be illustrated in a way that could be understood by others.

The angle of the hinge can be found by pointing a stick at the North Star. Looking at the world side-ways, a picture of a geocentric Universe can be drawn the most easily if east represents 'up' (rather than north we use today for 'up'). The rotation of the stars around the pole can then be drawn as a big circle to indicate the rotating sphere of the heavens.

An observer at Stonehenge who is looking east (represented as a dot on the top of the ball in the picture below) can see the North Star at about 39° anti-clockwise from a vertical line drawn straight up to the skies. The band of equatorial stars rotate at about 51° clockwise from vertical (the dashed line below) and the Sun also seems to orbit like the stars, but moves to about 24° above the equator in summer, and then moves down to about 24° below in winter:

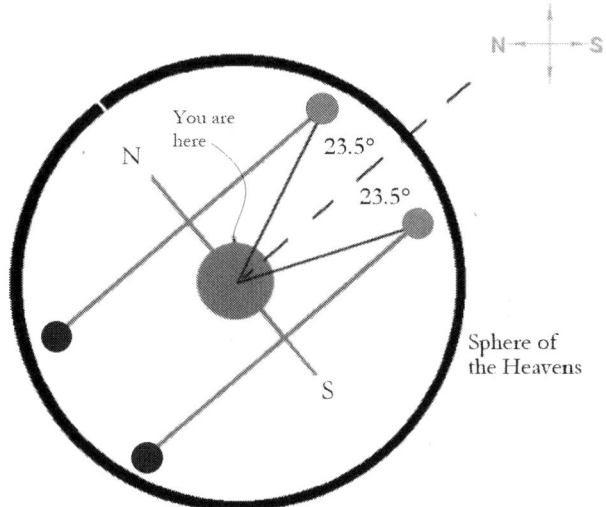

Drawing the Universe: England at the top of the world

The image above, *with an observer in England standing on the top of a ball*, shows one of the simplest geocentric explanations of our world, the heavens, and how they appear to move around us.

A vertical line drawn at Salisbury, England (a latitude of about 51°), is one seventh of a circle from the equator (51°/360°) and 6/56ths from the polar axis (39°/360°). So the circle of the stars, showing the blackness of the sphere beyond, can be neatly divided into fifty-six parts:

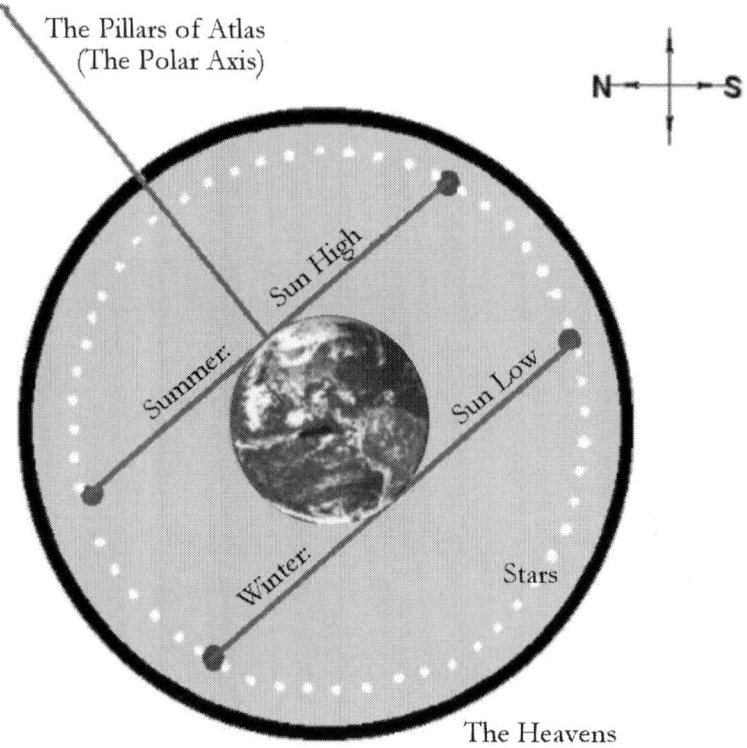

The 56 divisions of the heavens from Southern England (top).

This image (above) now duplicates Stonehenge's ground plan; with the stone circle representing the fixed ball of rock at the centre of the Universe:

5: STONEHENGE AND THE HINGE

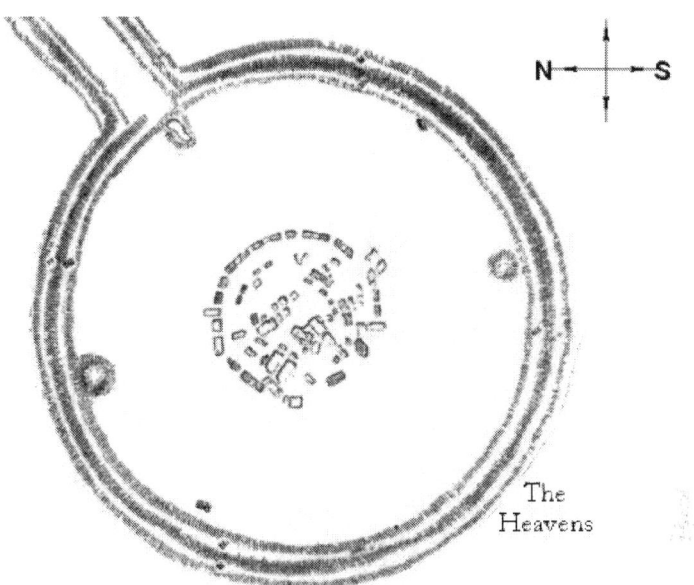

Modified & updated version of Stonehenge 1845 ground plan

But to draw the image above accurately, the direction of east, or some other cardinal axis, needs to be found. A very long straight pole could be pointed towards the North Star and a plumb-bob dropped to find north. However, even if a very long and perfectly straight tree trunk happens to be handy, it only works well when the North Star is easy to find.

As described earlier, another way to find the east-west line is to go up to high ground, or somewhere like Beachy Head, and to find equatorial stars. Provided that the star's rise, or set, can be seen, any of this band of equatorial stars, set close to what is known as the celestial equator, can then be used to find an east-west line. If the horizon can be seen in the distance, to either the east or west of a location, the line can be found by tracking one of the equatorial stars down to the horizon.

Stars seen rising and descending along the east-west axis can also be used to find the days when the Sun circles above the equator. On these days (known as equinox), two sticks aligned to sunset and sunrise will be almost perfectly aligned to east-west.

However, if the far horizon is behind hills, the Sun (or tracked stars) will set earlier than expected. The effect on setting out, if a west horizon is used to find east-west, is that the east-west line will appear to be rotated anti-clockwise from where it would have been if the far horizon could be seen. In other words, the method's east-west line would be slightly wrong.

The path of the setting sun

A way around this error is a method known as the Indian Circle:[02] A big circle is drawn around a vertical stick placed in an area of flat ground. When the Sun rises, the end of the stick's shadow will touch the circle. When it sets, the shadow will touch the circle again for a second time. A line drawn between the two gives the east-west direction.

This is a useful counter-check, but this method also has a potential problem: if the ground is slightly sloped (for instance a slope down from west to east), the early morning shadow, cast by the stick, will touch the western edge of the circle earlier than it would have, had the ground been completely flat. In the evening, with the western sun at a shallow angle, the shadow will touch the lowered eastern edge later than it should.

These two effects result in the same type of error which would be made when using a horizon which is raised by hills: if the ground slopes up to the west, the east-west line will appear to be slightly rotated anti-clockwise. If the ground slopes up to the east, the east-west line will appear to be rotated clockwise.

At a location like Stonehenge, where the eastern horizon is blocked by hills, the west, which appears to be flat to the horizon, makes an ideal choice to use to find out cardinal directions. The ground also appears to be very flat at this particular location, and this could also allow the 'Indian Circle' method to be used for checking that the cardinals are correct.

Unfortunately, these are both illusions: the horizon to the west is raised (there are hills in the distance; particularly directly to the west). These hills, for example Codford Hill and Clay Pit Hill, slightly raise the horizon above the true horizontal.

5: STONEHENGE AND THE HINGE

The ground at Stonehenge is also not as flat as it appears: it slopes down from west to east. At Stonehenge's latitude, if the sky was drawn with east representing 'up', a drawing of the heavens with 56 divisions of latitude would be rotated anti-clockwise by about two degrees or so:

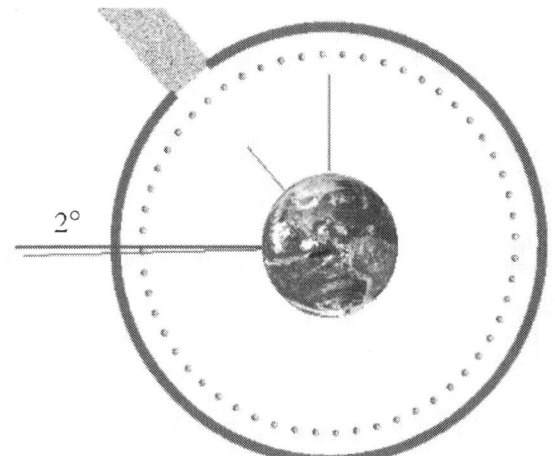

Setting out the heavens at Salisbury: East at top of picture

The original early features of Stonehenge, before stone was brought onto the site, become identical to the above setting out diagram: the 56 Aubrey holes at Stonehenge are rotated *anti-clockwise* from cardinals by a couple of degrees. A circular bank surrounds them. The Avenue extends out at about one degree *anti-clockwise* from 39° (the ideal polar axis line). The southern entry also appears to be slightly out, *anti-clockwise* from south.

If the plan of Stonehenge were intended to show the heavens, Stone hole 96 (the probable original position of the Heelstone) would also be slightly out of alignment (*anti-clockwise*). However, the Heelstone, which appears to have been purposefully moved to a 'new' position after everything else was constructed (now in Stone hole 97), is therefore in the right location to show the correct angle of the polar axis.

The original layout of Stonehenge therefore appears to fit a drawing of the world, and its Universe, when set out using simple, but very slightly incorrect, methods.

The Sun's apparent winter and summer orbits can also be laid onto an 'east=up' drawing and stone markers used to show the extent of the Sun's movement. Two orbit points would be needed to show both midday, and midnight, at the winter solstice; with two more for the summer solstice.

Apparent Orbits and Station Stones

These four orbit points, which define the Sun's movement, are in the same place as Stonehenge's four Station Stones (the drawing below shows the arrangement viewed from the west with east at the top):

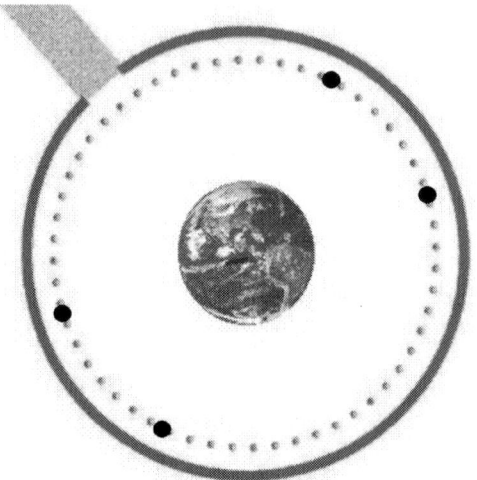

The Sun's orbits in a geocentric Universe: (East at top)

This representation now looks identical to Stonehenge's earlier original layout:

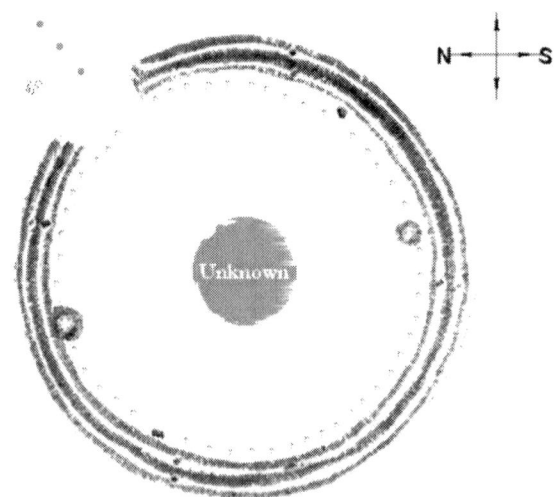

Modified version of Stonehenge 1845 Ground plan

5: STONEHENGE AND THE HINGE

The Sun appears to orbit the Earth whilst slowly moving by about 24° either side of the equatorial circle. This movement divides neatly into 30 parts (360/12), but does not divide into the 56 parts which would most easily describe the latitudes of the heavens in this part of the world.

A ring of 30 stones could instead be added to represent the Earth and to show the Sun's mathematical relationship to it. If this image is turned around by 90°, so that north is at the 'top', (the way we usually arrange maps today), this map of the Universe also becomes identical to Stonehenge's final layout:

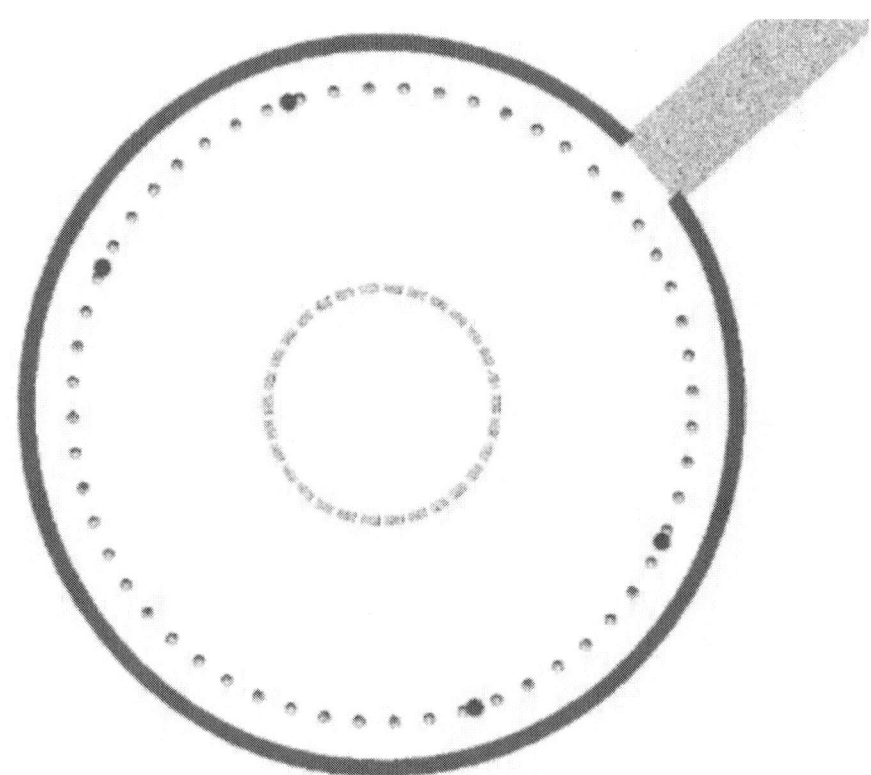

The Earth and solar markers: (North at top)

Accuracy of the design

If the previous method of drawing our Cosmos were part of Stonehenge's original design, the layout would originally have been one or two degrees out of alignment. Other than bringing a huge, perfectly straight, tree-pole into the area, lining it up with the North Star and dropping a plumb line down to the centre of the circle, there is little or no way to see if a mistake has been made. And the idea that a tree-pole would have been brought into Stonehenge seems somewhat unlikely.

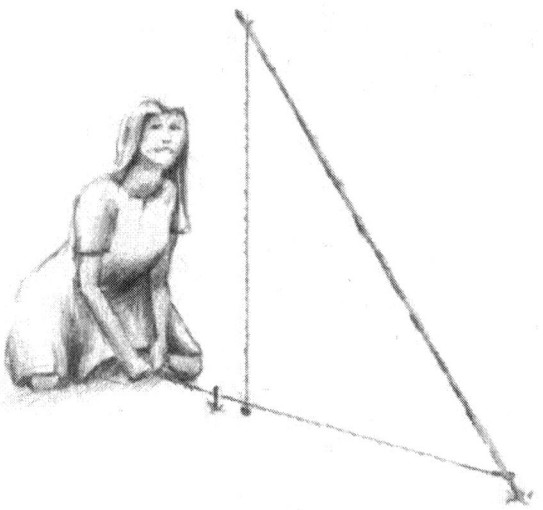

Finding north using a north-star pole & plumb-bob

But around about the time that Stonehenge was built, some of the Avenue's Stones seem to have been removed. The ones known to have been kept are the Heelstone and the Slaughter-stone. At about the time the sarsen ring was built, a ditch was dug around the Heelstone (after its counter-part stone had been removed), showing that the removal, or repositioning, of stones was intentional.

These two remaining, or possibly repositioned, stones are in the correct position to represent the real position of the polar axis relative to the true cardinal directions (north, south, east and west): if these two stones might represent a sudden increase in knowledge of how to find cardinal directions, the implication is that a more advanced method would have been used to set out the stones.

5: STONEHENGE AND THE HINGE

But when the horizon is elevated by hills and magnets are not available, the only good method to obtain a precise cardinal layout is to trace a line to the North Star.

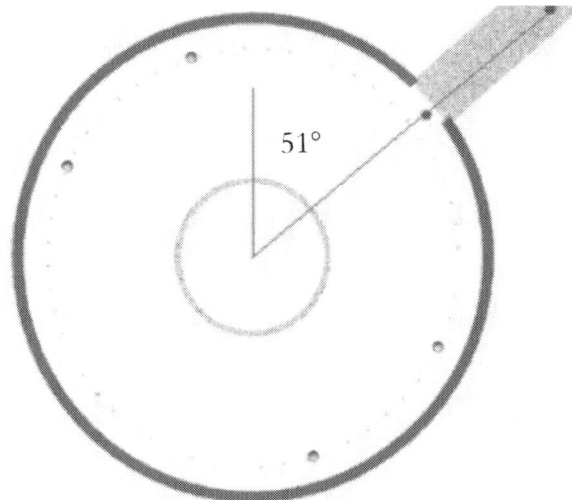

The Avenue stones which were kept (Heel & Slaughter Stones)

A straight tree pole aligned with the Pole Star, (Thuban at that time), could have been placed so that a plumb-bob is dropped to the centre of the circle. This might have been sufficiently accurate to show previous misalignments. But if a tree pole was brought to the centre of Stonehenge, why was it brought in?

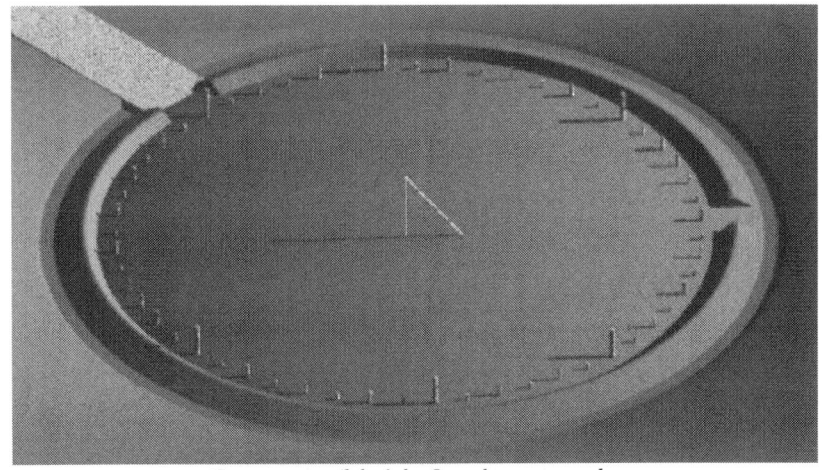

Computer model of the Stonehenge grounds

A summary of Stonehenge's layout

This chapter demonstrates that all of the plan layout components of Stonehenge's later phases are similar, or identical to, an early version of a geocentric cosmos. Its bank, layout of Station Stones, Avenue, external sarsen circle, and possibly the Aubrey Holes, correspond to this type of design layout.

If this was the purpose of the original layout, it might have been noticed, at about the time the stone circle was built, that slight improvements could be made. If that did happen, then those improvements also correspond to the features that were modified at this monument.

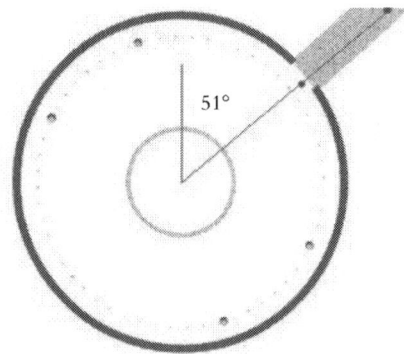

But the layout does not explain the construction of the central part of Stonehenge. And if a long, perfectly straight, tree pole was brought in to Stonehenge, why was it brought in?

6: A 'T' SHAPE DEVICE & THE COSMOS

The engineering device described later in this chapter (Chapter 6) was unknown prior to 2012. It needs no specialist materials and is quite simple to construct. However, it can be difficult to understand why it works. Publication of this sort of idea in a journal would be difficult, but I felt that a method of "peer review" would be helpful to show that it really would work.

At the time, the device was a) something that can be made or used, b) new and c) inventive (not just a simple modification to something that already exists). These criteria are also what is required to be granted a patent ("Utility grade": the highest form of patent). So I applied for a patent in 2010.

In 2018, a patent (GB2486636) for the device described in this chapter was granted by the British Intellectual Property Office.

The 'T' Rune in antiquity

The 'T' rune, 'Tir' or the 'Tiwas', which appears on weapons and crematory urns of the Anglo-Saxon period, may be references to the god Tiw [01] (or Týr in Old Norse), from whom we get the weekday 'Tuesday' (Tiw's day or 'Tiwesdæg' in Old English).

In the Anglo-Saxon Rune Poem, Tir is also associated with the Pole-star, in addition to, or perhaps rather than, a deity: [02,03]

> "Tir biþ tacna sum, healdeð trywa wel wiþ æpelingas; a biþ on færylde ofer nihta genipu, næfre swicep."
>
> "Tiw is a guiding star; well does it keep faith with princes; it is ever on its course over the mists of night and never fails."

Tyr by Lorenz Frølich*

The English language's sequence of days: Sunday (the Sun's day), Monday (the Moon's day) and Tuesday (the day of Tiw or the Pole-star) makes a logical progression if describing the sequence of importance of those things that exist in the skies. In Neolithic times, the Pole-star would have been Thuban rather than Polaris.

Saturday in Nordic languages was "Wash day" rather than a tribute to the Roman god Saturn, so probably described the day at the end of the week. In old Norwegian and Icelandic, the equivalent rune Týr refers to a one-handed god; which also fits well with the idea of referencing the pole-star and the axis around which the heavens rotate.

* *Image of Tyr by Lorenz Frølich extracted from:*
https://commons.wikimedia.org/wiki/File:T%C3%BDr_by_Fr%C3%B8lich.jpg

A device which could be used to find the North Star, a stick pointing to the North Pole, can also be used with a sail-stick so that the sail traces a non-equatorial star (or even our own Sun). The image below shows a stick arranged, at any place with a latitude of 51°, so that it can be rotated to point towards the Sun. During high summer, when the Sun is located approximately 24° north of the equatorial band of stars, the stick must also be offset by an angle of 24° from the plane of the rotation:

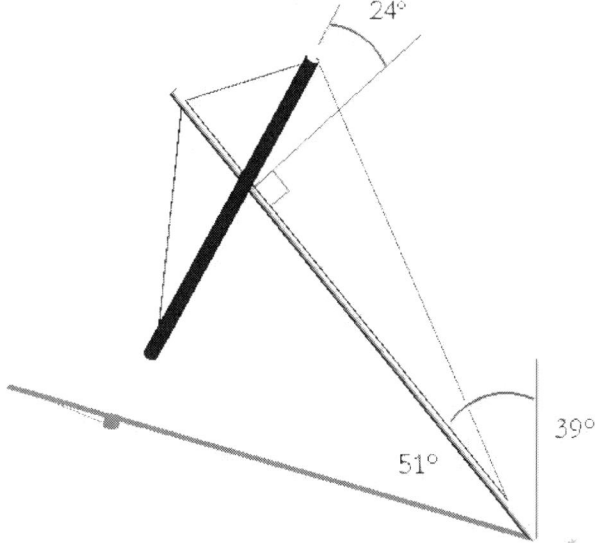

To trace the summer sun on a geocentric world

Whilst there is no evidence that a tree was brought into Stonehenge, there is a socket at the base of Stone 54; which is the right size and shape to fit the end of a tree-trunk. The low socket on Stone 54 is also directly south of the centre of the monument and is in exactly the correct location to work. As described in the earlier "Potted History" section, this stone is the heaviest at Stonehenge and has an enlarged foot below ground level.

However, this large socket seems exceptionally well-worn; almost as if it had been damaged by use. This suggests that it was not some sort of temporary works used for the construction of the monument (although the damage might also be due to Victorian souvenir hunters).

The principle of a polar axis at 51° from horizontal, together with a rotating arm to track stars, was used at the Greenwich Observatory to create what is known as the 'Equatorial Group' of telescopes. When these telescopes are at right angles to the support axis, they can track equatorial stars. If the angle is changed, other stellar objects can be tracked:

An equatorial telescope at Herstmonceux, Sussex

The same idea can also be used to demonstrate how the Sun appears to rotate around our world. From the top of a ball, with the topmost surface always representing Southern England, the reason why the Sun appears to disappear from view during night-time can be shown:

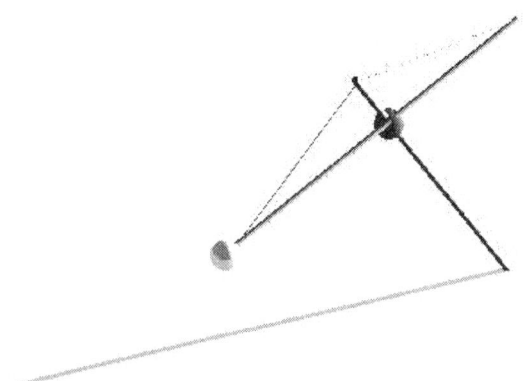

Pole and stick arrangement to show the sun rotating around a geocentric world

6: A 'T' SHAPE DEVICE & THE COSMOS

Before Stonehenge was conceived, it seems that tin and copper had been discovered.

Tin is a material which reflects light. It can also be easily moulded to shape using an ordinary fire.

After casting, it can be polished using clay and/or chalk paste (powder combined with water or oil) to make a mirror.

Cast tin after cooling

Cast tin after hand polishing using chalk paste

Tin degrades if it is kept in a cold climate (a process known as 'tin pest'). Unlike almost all other metals, it gradually turns to dust over centuries; giving no evidence that it had ever been used. Unfortunately, this effect means that the discovery of tin mirrors in archaeological digs would be exceptionally unlikely.

Crude tin mirrors have a reflectance greater than 50% and can easily be used to reflect a bright light onto a ball. The idea of using reflected light could then be used to improve any demonstration which shows how the Sun seems to rotate around the Earth:

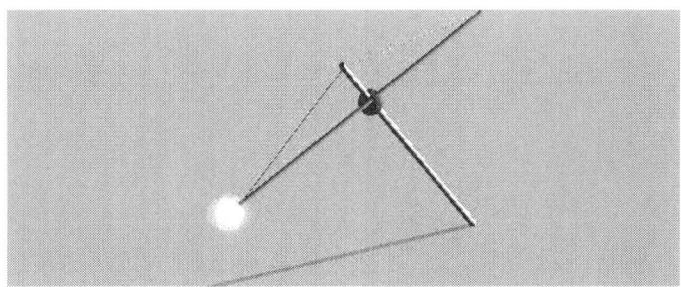
Using mirrors to light up a model of a geocentric Sun

A problem with using reflections to light up a sun-ball is that the mirrors have to be constantly re-focused, to keep track of reflections, as the Sun moves:

Adjusting the mirrors

However, if the mirrors are arranged as a sphere and the sail rotated, (rather than moving the mirrors), the mirrors will remain automatically focused for several hours providing the sail is set at the correct length (about half-way between the centre and the face of the mirrors):

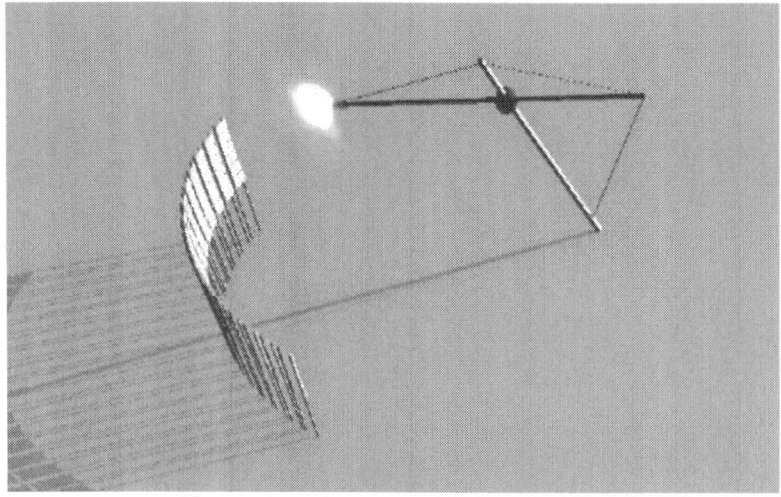

Using spherical mirrors to make a focal device

Spherical concentration is used at the Arecibo Observatory,[04] where a giant spherical collector has been arranged to collect radio waves from distant galaxies.

When used in a demonstration device, sets of spherical mirrors have to be kept tightly in place so that they do not move in the wind. An inwardly facing cylinder, a circle of very strong material, is therefore required at high level together with flat faces against which the mirrors can be measured and tied.

To ensure that the mirror sets form a sphere, which has the same centre for all the mirrors, this circle of strong material also needs a perfectly level rim:

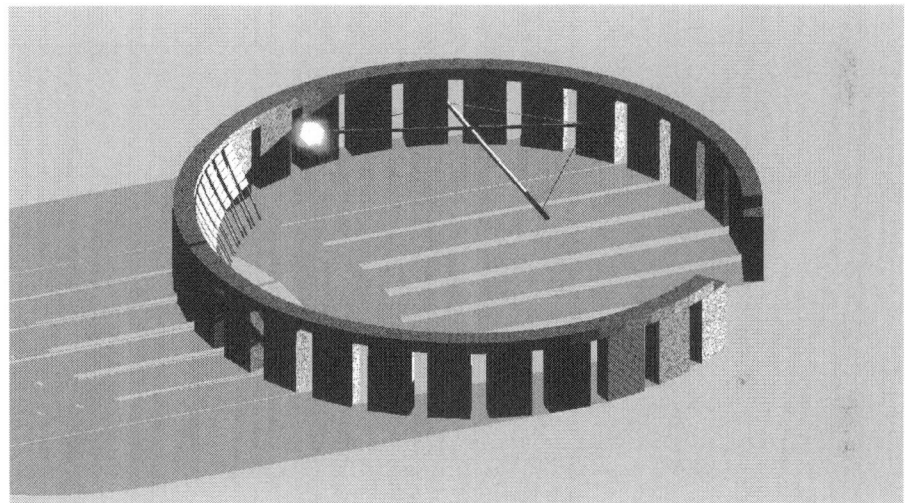

The setting ring: To support winter mirrors

At Stonehenge there is a high-level ring of inwardly facing lintels arranged as a shallow cylinder. Due south of the centre of this circle, a stone pair (stones 53 and 54) have a socket into which a tree could be inserted (Stone 54 has an unusually enlarged foundation). If the tree were pointed at the North Star, and a second pole rotated around it, mirrors placed against the lintel rim would focus light to a ball.

As the Sun rises in its circle, the ball descends along the path of a circle. As the Sun sets, the ball rises. If a demonstration were needed of how the Sun moves in a Universe believed to be geocentric, this is the perfect arrangement.

However, tin and copper were newly discovered materials. They were probably more valuable than gold is today and would not be left out overnight. Metals would need to be stored in a safe place, moved to location in the morning, and then set up so that they focus.

This arrangement would therefore be ideal for the afternoon (when the ball appears to be rising). To reflect, mirrors need to face towards the Sun and, in Southern England, the Sun travels from south to west over an afternoon. Therefore mirrors must be placed in the north-east; with their reflections bounced up onto a ball which would be seen to slowly rise towards the south within the north-east part of the arrangement.

The best place to see this effect would be from the north-east. A wide avenue, such as the Avenue north-east of Stonehenge, would be ideal to show the effect of a bright mini-sun to a huge number of people.

The three season device

The device described so far has a rod that points up by 24° in the winter. This allows the shining ball to appear just above the top of the supports. However, the arrangement described is only good for winter.

After winter passes, the Sun gets higher in the sky. The rotating rod must point at the Sun, so the reflector must point downwards. At equinox (spring or autumn), the rod is at right angles to the pole. In summer, the Sun is high, so the reflector-ball would be even lower.

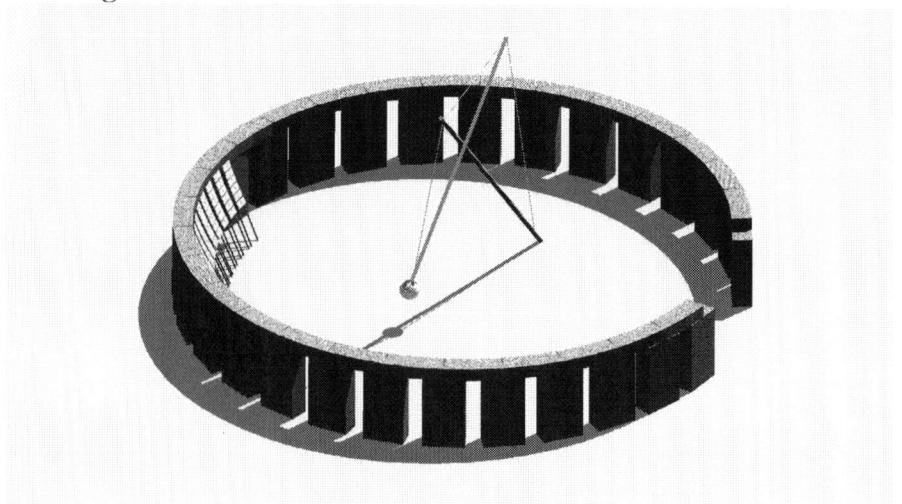

Summer: The sail pointing down

The effect of this, after winter passes, would be that the focal point (on the ball) would disappear below the top of the mirrors when viewed from the outside.

One way to improve this design is to raise the axis by just enough so that the whole assembly is again visible above the ring. As summer approaches, the whole assembly can be raised for a second time:

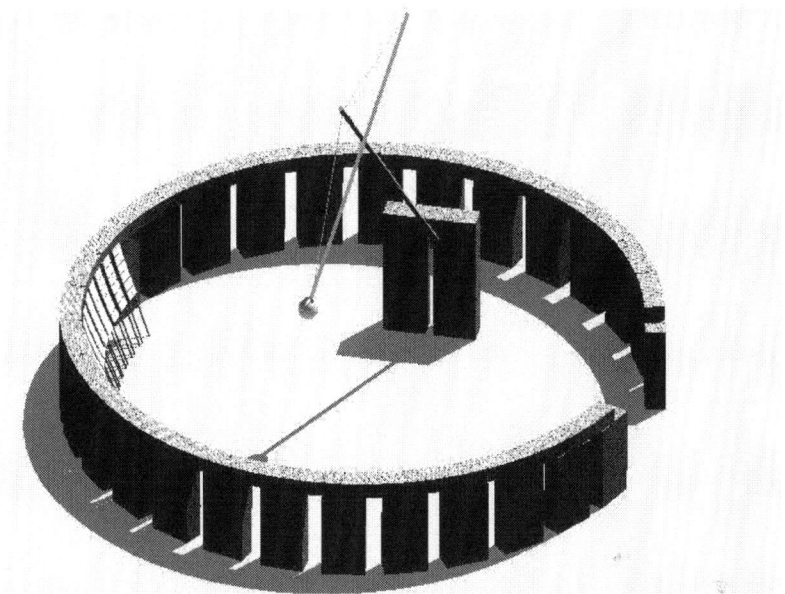

High level position of the 3-season socket arrangement

When dimensions are calculated, it proves to be possible to position the pole so that the ball of light is always in the correct position above the rim of circular stone.

A pole is very light by comparison to the heaviest stone (stone 54) at Stonehenge.[05] It can be lifted and hauled to its final position, a little like a May Pole, using the counter-balance of the stone's weight:

Mid May; the first day of summer: Raising the Maypole

And the pole position and angle set using a plumb-bob and marker; located beforehand using perhaps a smaller pole set to the North Star:

Setting the angle and location

To make this work at Stonehenge, two extra sockets would be needed so that the assembly can be moved up.

At Stonehenge, two extra sockets appear to exist on stone 54. These are in the positions required to allow the shining light to rise above the lintels in any season. The holes also appear to be the right shape and orientation and are located in the stone which has the enlarged foundation:

Stones 53 and 54: Stonehenge

6: A 'T' SHAPE DEVICE & THE COSMOS

A second effect of moving the assembly up is that the mirrors must be moved over and then tilted so that the spherical centre moves up with the apparatus. In summer, the mirrors are highly angled so are easy to prop from the ground. But at equinox, a second set of low-level supports would be best introduced, just inside the main ring, to prop the tilted and shifted mirrors:

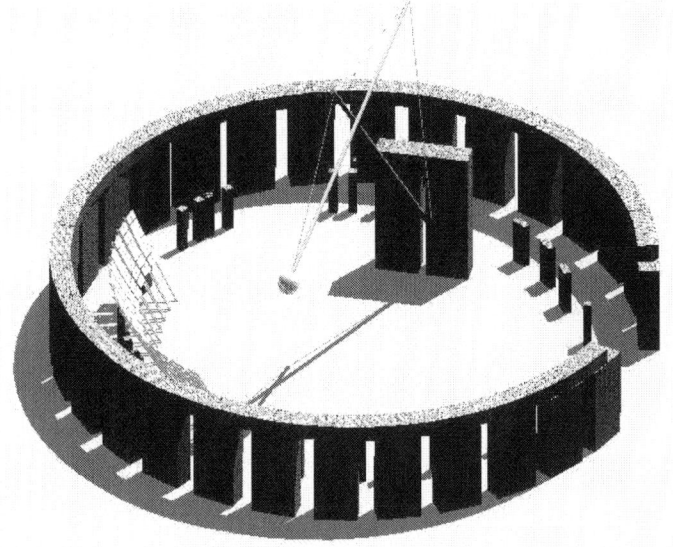

The equinox support requirement; Bowl centre raised by 2 metres

At Stonehenge, a second ring of stones, known as the outer bluestone circle, exists in just the right place to provide solid support points for mirrors which are tilted and shifted for spring or autumn:

The outer bluestone circle

The rotating collector is heavier at one end than it is the other. To improve the design, either a counterweight can be hung off the end, or the end can be tied down to a series of holding down posts:

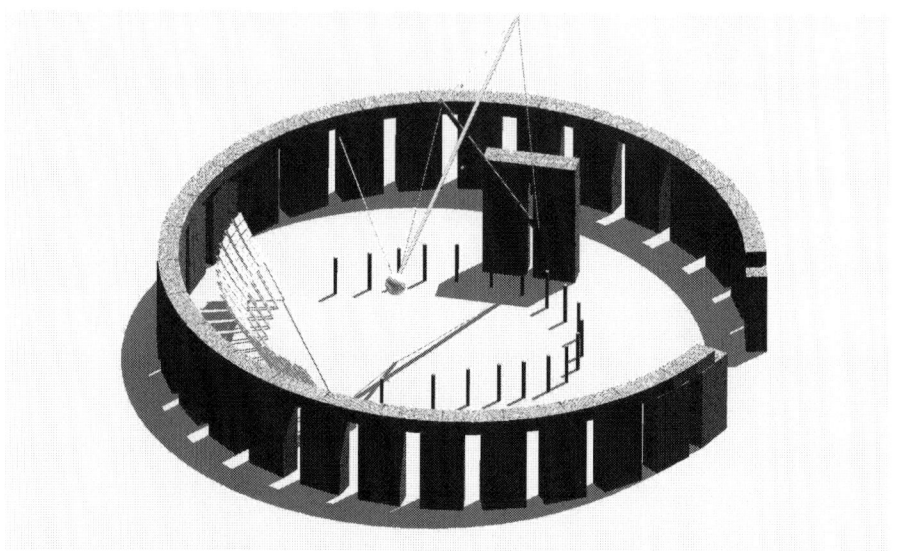

Oval or horse-shoe counterweight ring

At Stonehenge, a set of bluestones forms a circular corridor. This corridor happens to be suited to allowing the high end of the rotating pole to be guided or, possibly, tied down. These are not essential to a 'geocentric demonstrator' design, but these stones happen to be placed in a location that would not obstruct the operation of the device:

Stonehenge's inner bluestones

6: A 'T' SHAPE DEVICE & THE COSMOS

To allow the rotator to be hauled around the polar axis and also firmly tied against the wind, tall and strong platforms are needed in the north-east, arranged so that they do not cast too much shadow on the mirrors. To enable the fitting of a reflector, the end of the sail would need to be accessible from these platforms:

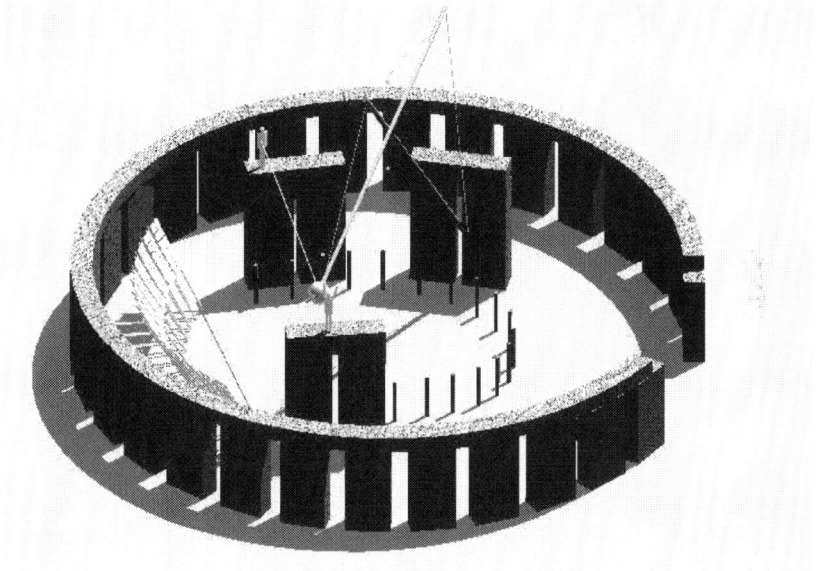

Adding the loading and rotation platforms

At Stonehenge, two sets of trilithons (Stones 51&52 and Stones 59&60) are set in just the right position to allow the sail to be rotated, be restrained against the occasional gust of wind, give access to the reflector and to not cast too much shadow on the north-east mirrors.

However, when the pole is raised by one level for either the spring or autumn, the end of the sail is only just reachable (so that it can be fitted with a reflector). In summer, with the pole two levels up, the reflector is far too high to allow access to the end of the rotating arm.

A way to get around this is to install two extra sets of platforms, in the south-west, so that the end of the sail can be reached no matter which position the pole is placed in. Because these must be higher platforms than the two used at the north-east, they need to be set at the back so that they cast the least amount of shadow:

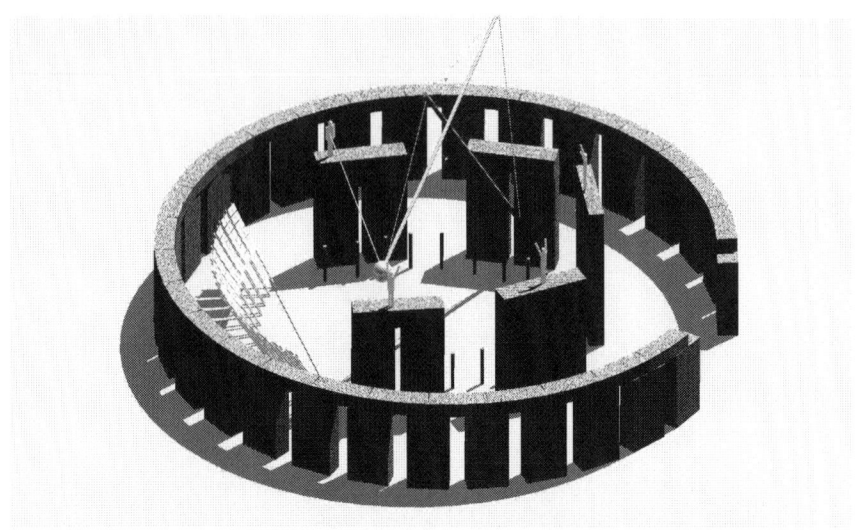

Adding platforms for equinox and summer

At Stonehenge, two sets of trilithons (Stones 55&56 and Stones 57&58) exist at just the right position and height to allow these operations to occur.

The three season device would create a ball of light. That ball would appear to glide on a sunbeam because its sail always points towards the Sun. As clouds appear, the ball would turn dull and with each new ray of sunshine it would light up:

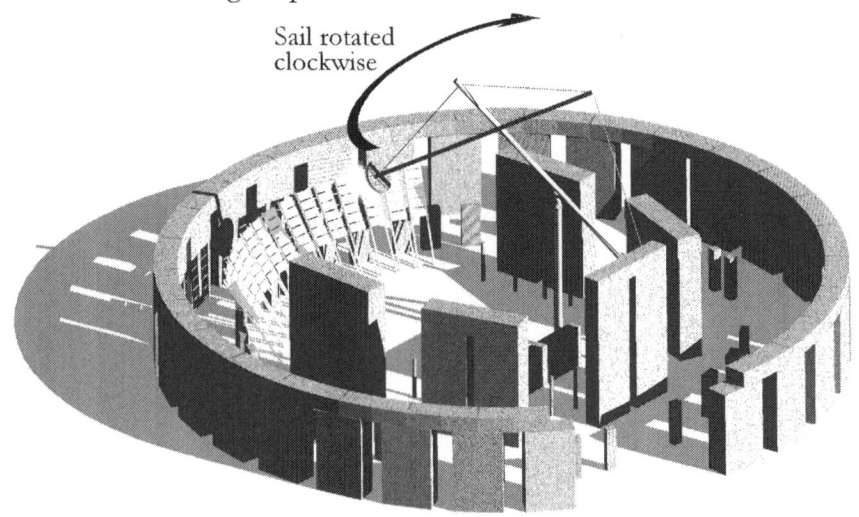

Rotation of the sail

The reflector can be a number of shapes. It can also be made to concentrate light in certain directions, but the overall effect is to create a small version of the Sun which appears to rise on the sail towards the heavens.

View from the north-east

Pre-testing

If such a three-season device had been installed at Stone 53/54, the position would need to have been checked before the stone were put in place: a mistake could make the arrangement unusable.

A pre-test of the highest point of the device, without a stone socket to secure it, would need a long pole set into some form of mound; perhaps four or five metres south of Stone 53/54.

Evidence of that type of mound exists at one location: immediately south of Stones 53/54 at Stonehenge.[06] This is the location that would have been required for a position check prior to erecting the stones.

The 'T' Rune

In late 2012, English Heritage published an archaeological report entitled 'Stonehenge Laser Scan: Archaeological Investigation Report'; which contains many newly discovered details. Both of these publications contain fascinating new information about the lives of, and the monuments built by, the people who constructed Stonehenge.

The latest scans have shown that there are over 100 images of a T shape carved into the faces of the stones at Stonehenge:

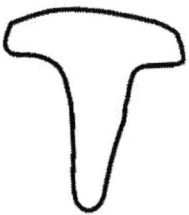

An example of the type of engraving found at Stonehenge

This type of engraving, thought to be an axe-head, does not occur to any extent anywhere else in the British Isles. The 2012 report shows the 'axe-heads' to be concentrated on and around Stones 3, 4, 5, and Stone 53. These stones reflect more non-symmetrical features of Stonehenge: Stone 53 is directly south of the centre and the other two stones are directly east.

Pages 28 to 31 of the report show the majority of newly discovered markings to be on Stone 4, *the eastern stone*, with other 'T' shape markings existing on stones 3 and 5, either side of the eastern stone. Apart from these stones, discussed in more detail in a later chapter, few other engravings exist. Page 34 of the report shows that the majority of the other newly discovered 'T' shaped markings are on Stone 53.

For the proposed device, stones 53 and 54 would be the most important part of the arrangement.

Large 'T' shapes have also been discovered on the western side face of Stone 54, the eastern side face of Stone 53 and the rear of both 53 & 54:

View of Stone 54 showing the 'T' shape
Photograph courtesy of Terence Meaden ©2013

If these shapes are not natural (or just an unintended consequence of working of the stone), the 'T' shape was particularly important at Stonehenge's Stones 53 and 54: where they appear to be concentrated.

The 2012 archaeological report[07] also shows more non-symmetry at the south point of the horse-shoe: Stone 54. This is the southern-most stone of the inner monument and is both the heaviest stone and the most notably unusual in comparison to others nearby:[08]

> "Though it is notable how well Stones 53 and 154 are finished in relation to Stone 54"

The report also states that the stones had been shaped to provide a view from the Avenue (in the north-east):[09,10]

> "The most regular and extensively dressed stones are however placed towards the NE"
>
> "This confirms the view that the sarsen circle was not supposed to be approached, or perhaps even seen, from this southwest direction (Tilley et al. 2007). Instead the emphasis was on the approach from the NE"

And that stones had been designed *not* to be viewed from other angles. It states that the surface finish of the stones becomes progressively less good towards the south west:[11-13]

> "The use of small stones (e.g Stone 11 and Stone 21) and irregular stones (e.g. Stone 14) is a distinct feature of the SW half of the monument"
>
> "The absence of working on the exterior faces of Stones 14-16 and coarse finished on the backs of Stones 10 and 11 is significant as these further demonstrate a significant contrast between the NE and SW sides of the monument, as well as indicating interior/exterior differences"
>
> "The absence of dressing on the exterior surfaces of stones on the SW of the Sarsen Circle indicates that the monument was not designed to be approached from this direction"

The report also shows that the lintels to the north-east are the most extensively worked:[14]

> "The lintels have been extensively shaped, with curved outer and inner faces matching the circumference of the monument"
>
> "Lintels 102 and 130, 101 and 102 on the NE half of the monument, are most regular lintels"

These factors all fit exactly with the idea of its use as a geocentric device.

Decay

Most modern structures are facilities which are designed to contain or carry other things. For example, a museum's purpose is related to its contents; but the contents will always be the first thing to be removed when the building becomes outdated or threatened.

A solar structure, of the type described here, contains two sets of temporary features. The first set of temporary features are the mirrors and the reflector, both containing metal; a precious commodity. Additional temporary structures (such as the pole, ladders, ramps, and so on) form the second set.

At the time Stonehenge was constructed, the metal components would have been more valuable than a structure such as Stonehenge. Containing and using metal would have been the primary purpose of a building designed to be a solar concentrator. As with a museum, these valuable contents would be the first to be removed if the structure either fell into disuse or came under the threat of pillage. In the unlikely event that any metal was left behind, tin would gradually crumble to dust over the centuries; leaving no trace other than an elevated tin content within the soil.

The second set of temporary features (ladders, ropes, poles, platforms, and so on) might be left in place in the event of a sudden fall into disuse. Because these items are less valuable, they could also be stored nearby. Either way, timber left above ground eventually rots and thus returns to the soil leaving no trace.

In other words, the structure described in this introduction would leave no trace other than the structure which currently exists at Stonehenge. However, tokens representing the structure of the cross within a circle, if produced in gold, may have survived:

A gold sun-disc [15]

Summary

When the Sun leaves the far north in the time of winter, the land becomes cold. On a world believed to be geocentric, the Sun would also be seen to turn within the celestial sphere but would appear to have a will of its own. Knowledge of how the heavens work could therefore have been seen as fundamental to the continued existence of people living at the edges of the world.

Stonehenge's plan layout can be shown to be the same as an idealised geocentric description of the Universe. Its inner stone monument is demonstrated here to be capable of producing a spectacular public display of solar movement. The arrangement of this system is shown to be based on a simple method of tracking celestial objects.

Therefore, Stonehenge could have been both a depository of knowledge about the Universe and a place of learning designed for popular interest.

7: THE SIZE OF THE WORLD

In the earlier chapters, Stonehenge's layout was shown to be the same as an idealised geocentric description of the Universe; its inner stone monument capable of producing a spectacular public display of solar movement using reflected light. But Stonehenge is a single monument; its apparent arrangement (as a place to teach knowledge of the Universe) could be entirely coincidental. In Northern Europe, the Druids considered it unlawful to commit their knowledge to writing,[01] so there are no known records of what might have come before.

But if knowledge of the world and the heavens were well known, there should be traces in the far past, well before Stonehenge, of the methods that might have been used to find that knowledge.

- → Knowledge of how the Sun and the heavens revolve;
- → The ideas used to explore what the world might be;
- → The systems used to prove that knowledge;
- → The methods used to explain what was being done.

At some stage in the past, nobody would have known for sure what type of world they were living in. They might have had a good idea of what the Earth is, but proving those ideas true would be difficult. This chapter looks at how the shape and size of the world might be found using the landscape available in Britain.

It also shows a simple method that would produce earthworks that are remarkably similar to the constructions of the Neolithic period.

History

In the early centuries BC, Eratosthenes[02] proposed a method using the angles from shadows, measured downwards from the Sun. Those angles will be different at other places in the world and, by measuring the north-south distance between those places, the world's size can be calculated.

In the medieval Islamic era, Al-burini, one of the greatest scholars of the age, used a method[03] which relies on the angle down to the horizon from a (known) height of a mountain. Special equipment is needed to measure the angle down, and a simple calculation allows the world's size to be found.

Even simpler variants of Al-burini's method exist. For example, if a hill looks over sea in both directions, special equipment (to measure the angle from the horizontal down to the horizon) is no longer required: all that is needed is to find the difference between a) the angle down to the horizon in one direction and b) the angle down in the other.

When I started looking for locations to show this, the idea was just to prove that these things could have been done and not look for signs that they may have already been done. I selected locations using topographical maps and then carried surveying kit up along the paths of the South Downs to test the ideas. However, on arrival, I found Neolithic monuments all arranged as if designed for the purpose. The surveying equipment needed (for Al-burini's method) was unnecessary.

In this chapter, monuments in East Sussex are used to show that inquiry into the nature of the Universe could easily have produced a science based on the idea of a geocentric world. The photographs that follow are of real places and tests, but similar monuments also exist elsewhere.

7: THE SIZE OF THE WORLD

The idea

On a curved world, it should be possible to see the world's curvature from a hill by finding the slope down to the horizon. On a very tall mountain, the angle down to the horizon will be easy to see:

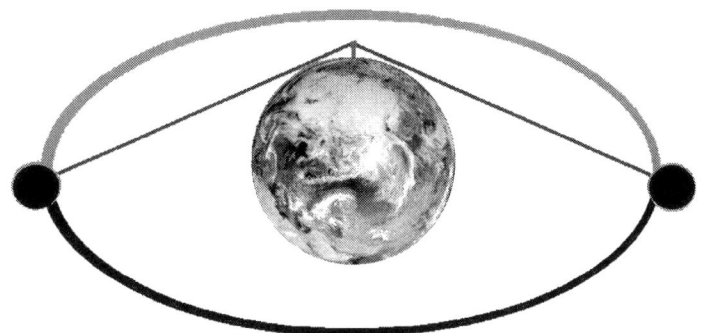

Seeing the curvature of the Earth using the Sun

If the Earth seems to be a ball, it is possible to work out its size using only wood, the land, and a variant of Al-burini's method:

At sunrise or sunset, on a clear day, it is possible to be sure that the horizon is not obscured by haze at that moment when the Sun peeks above the horizon. Twice per year, the Sun rises directly in the east and then sets directly in the west. At this time of year (equinox), the difference between the angle at sunrise and sunset can be measured using two sticks:

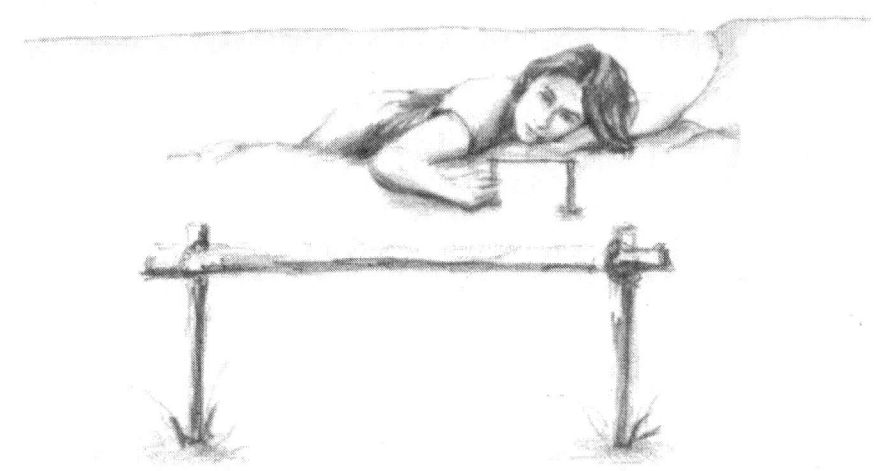

Using sticks as horizon sight-lines

At small islands, or places where the land juts out into the sea, the discovery that the horizon slopes down in both directions could have been found by chance. For example, in south-eastern England, looking over sea views at the far tip of Beachy Head, there is a ridge, where Neolithic barrows once existed,[04] at which this feature of our world can easily be seen:

The highest hilltop ridge adjacent to Beachy Head

This small ridge is some 165 metres above sea level and adjacent to the cliffs of Beachy Head. It is also adjacent to the ancient track known as the South Downs Way.

Using the rates of cliff erosion below the nearby Bell Tout lighthouse, the cliff would have been some 400 metres or so further south in Neolithic times.[05] Nevertheless, this ridge, with views to both the east and west, could be used to find the slope down to the horizon thousands of years ago.

Finding a hill's height

Next to cliffs of Beachy Head, the easiest way to find the height above the sea is to drop a rope down. The geological bedding of the chalk is horizontal and this produces cliffs which overhang in places; allowing a very accurate measurement down to sea level:

The cliffs near Beachy Head

Another method to find the height of a hill is to use a level, for instance; a water-filled trough cut into timber, to sight onto a pole stood on lower ground. The height between one spot and another can then be found by sighting horizontally over the water in the trough.

A simple trough level is not very accurate compared to modern instruments, but could work well over short distances:

Using a trough as a level

This process can be repeated in short distances, all the way down or up a hill, and each time adding the difference in height to get a total hill height.

This second method, using a trough-level, could also be used near Beachy Head, where the seafront is just over a kilometre away; along the eastern paths which gently lead down to Eastbourne's seafront:

Looking east from Beachy Head

Looking North towards the South Downs from Beachy Head, there are several other hills which are higher and could give better accuracy for any experiments that might have been done to find the size of the world. For example, Willingdon Hill's brow is at 184 metres above the sea, Bourne Hill is up at 202m and Firle Beacon is at 217m.

Each of these has a large Neolithic barrow at its summit:[06]

The hills to the north of Beachy Head

7: THE SIZE OF THE WORLD

Bourne Hill, the nearest summit, is also sited next to the South Downs Way and has a very large flat-topped tumulus which looks both east and west over sea views:

Bourne Hill Tumulus

One method of finding the difference in height between two points is to sight horizontally to the top of a pole and to drop a measuring line down. The same system of sighting to the next pole down can then be used downhill and over long distances. Each drop in height is added to the last and, when the sea is reached, the height of the hill is the total.

In a location such as the eastern hills of the South Downs, when at low level, the level can be easily checked against the sea's horizon. This method could potentially give an accuracy as good as 0.02°, depending on the eyesight of the user.

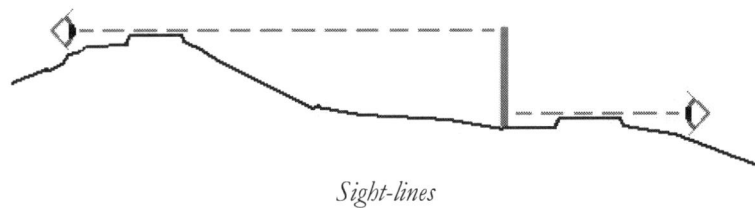
Sight-lines

A shaved pine pole, perhaps 100mm diameter (weighing about 40kg) can be carried by two people. Anything bigger than this, though difficult to carry, could be used but on a more permanent basis.

Once lifted to an angle, a pole can be hauled up to position using rope and then tied to stakes. To get any higher than 12 metres probably needs a second thinner pole, hauled up to position using rope rather like a sail.

Although it is possible that trees could have grown taller in Neolithic times, trees today rarely grow to greater than about 25 metres in the nearby Friston Forest. This puts a limitation of about 20 metres, probably less, on the height difference that can be found between any two locations used to measure the drop.

From Bourne Hill (202m above sea), due south to the hills by Beachy Head, the shortest route is along the high ridge of the ancient track known as the South Downs Way. Along this track-way, there are three intervening hill brows with a maximum distance between each of about 1.5 kilometres. These three exist at the high points of Foxholes Brow (about 183m), Beachy Brow (about 168m) and Pashley Hill (about 167m).

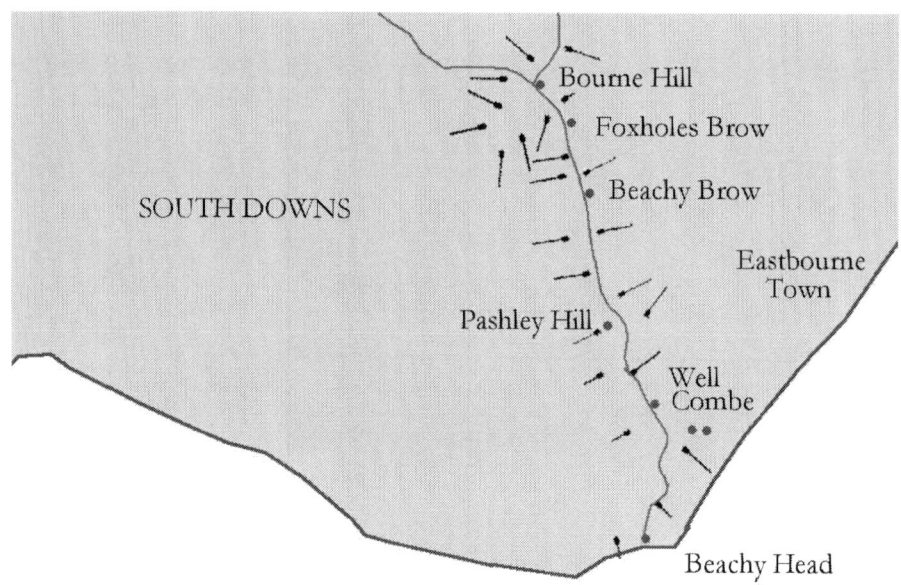

The route from Beachy Head to Bourne Hill

On each of these high points there exists at least one Neolithic bowl tumulus; typically about 8 metres diameter. Each of these is at a high point location which has direct views of the sea, the next hill, and other smaller tumuli in the series.

Foxholes Brow Tumulus:

Foxholes Brow bowl tumulus (one of two tumuli)

View of Bourne Hill from Foxholes Brow looking north

Beachy Brow Tumulus:

Beachy Brow bowl tumulus

View of Bourne Hill & Foxholes Brow from Beachy Brow

Each of these tumuli is ideally sized and located to provide an accurate sighting platform to the next tumulus (or anything else) along the South Downs Way.

From doing the experiment myself, I found that a raised platform was very helpful to prevent shrubs and trees from obscuring the view. In addition, any sort of experiment is helped if the top of the platform is flat. This reduces the chance of gusts of wind affecting the experiment.

Sighting on a flat bowl

From Bourne Hill, smaller Tumuli run all the way to Pashley Hill, where another sight-line goes to a Tumulus at Well Combe, followed by two more tumuli, each about 15-20 metres or so below the next. According to English Heritage, and about 15-20 metres below the lowest of these two tumuli, the ploughed out remains of yet another possible tumulus (Monument 152278) were once set within an area which was later to become a prehistoric field system.

The seven or eight smaller tumuli described in this part are just a small sample along this route. [07] For example, between Bourne and Foxholes, there are three more small tumuli, all arranged to have sight of others. Between Pashley Hill and Beachy Brow, there are another half dozen. All of these tumuli are along, or are set very slightly away from, the South Downs Way.

7: THE SIZE OF THE WORLD

The remains of tumuli along the South Downs Way, between Beachy Head and Bourne Hill, are all located in a way which would allow the height of Bourne Hill, or other locations, to be found with reasonable accuracy.

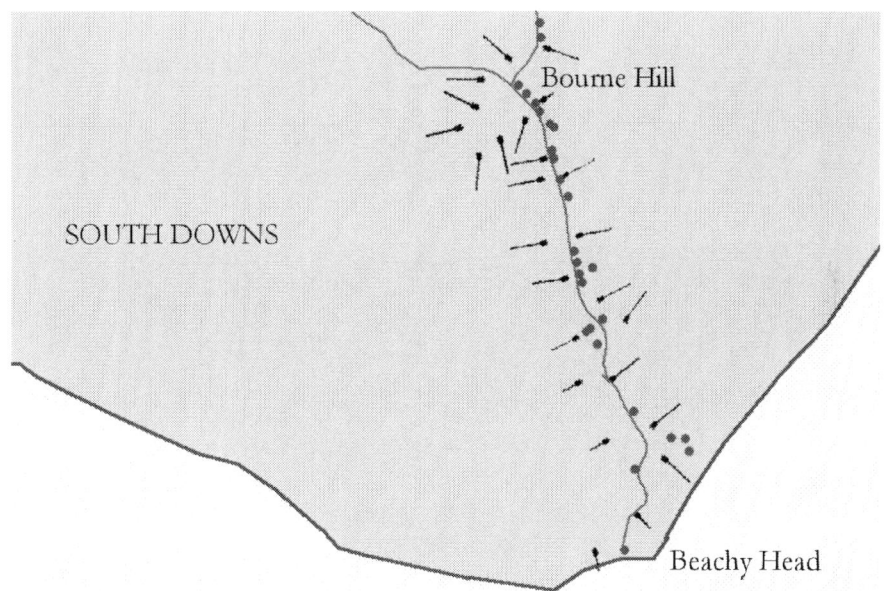

Location of tumuli near Eastbourns (between Bourne Hill and Beachy Head)

A modern experiment

It is possible to line up a horizontal bar so that it aligns to the horizon as seen when the Sun rises at the equinox. To test the theory, I ran the experiment on Bourne Hill: two sticks were adjusted so that they met with the sunset at the moment the Sun disappeared from the horizon:

After sunrise looking west

When the Sun rises, the sticks are not on a level with the horizon as seen in the other direction. The height difference was then measured:

Bourne Hill: Sunrise at equinox

In the above rough experiment at Bourne Hill (autumn equinox 2012), the measured height difference was just under 150mm (6") over a distance of just under 10 metres (about 32') giving a measured slope of about 1:130 in each direction.

Calculating the size of the world

The radius of a circle (R), measured using a shallow slope, can be approximated by the calculation: **2 x h / (Ø x Ø)**, where 'h' is the height above the circle (the height of the hill) and Ø is the slope. If the slope denominator is an integer 'N'[08] (for example 1:100), the calculation simplifies to **R = 2 x h x N x N**

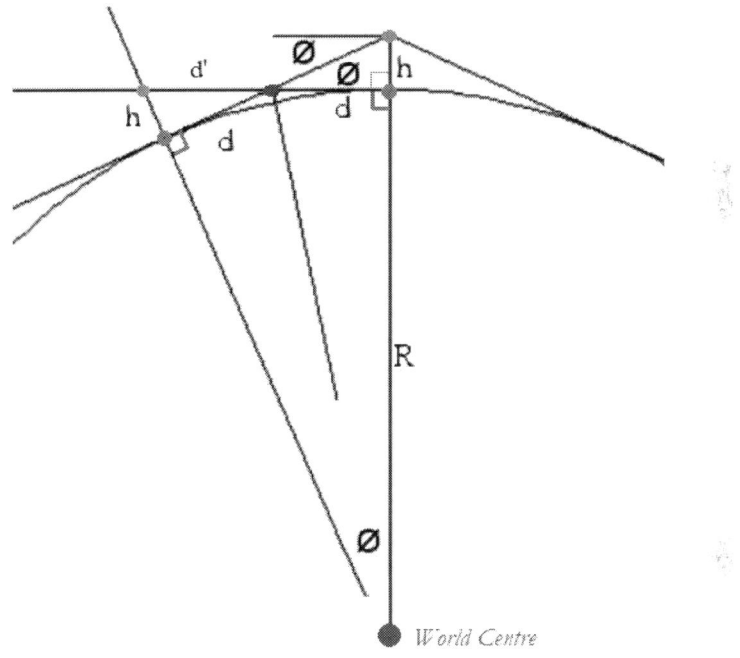

The angles needed for slope approximation of the size of the world

The approximate radius of the Earth (as measured using this rough experiment in June 2012) was within a few percentage points of the real size of our world.

A person with younger and sharper eyes than me should be able to see even better angular resolutions, especially if the whole width of the tumulus was used.

Describing how to calculate the size of the world

From Bourne Hill, another sight-line exists to a hill named Cold Crouch, a short walk away along a northern fork of the South Downs Way. On top of this high point (approx 182m) is a further Neolithic bowl tumulus. Between are two further tumuli on an intervening ridge named Babylon Down.

To the immediate north-west of Cold Crouch, a singular unusually tall bowl tumulus exists at the summit of Combe Hill (193m):

Combe Hill: Tumulus A

This location, like all the other bowl tumuli in the area, has good sea views to the east and to the west. A good way to visualise how the 'slope' calculation works is to imagine being on the world, and seeing the Sun's angle down from horizontal in both the morning and afternoon:

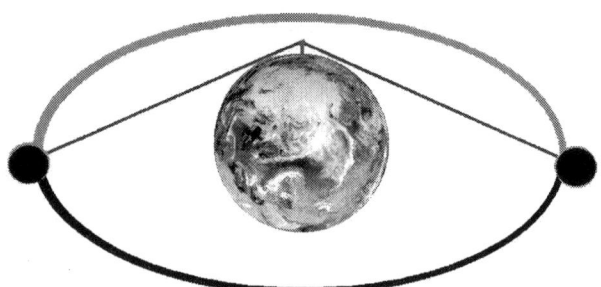

Seeing the curvature of the Earth using the Sun

If the above drawing is done to a large enough scale, and with large angles to demonstrate how it works, a person could sight onto a 'sunrise point', from a giant ball drawn on the ground, and check that the size of the world really can be found by using twice the height divided by the square of the slope.

7: THE SIZE OF THE WORLD

Immediately north of Borne Hill, at a place known as the Combe Hill Camp, there is a one-off Neolithic structure which is slightly larger in scale than Stonehenge:[09]

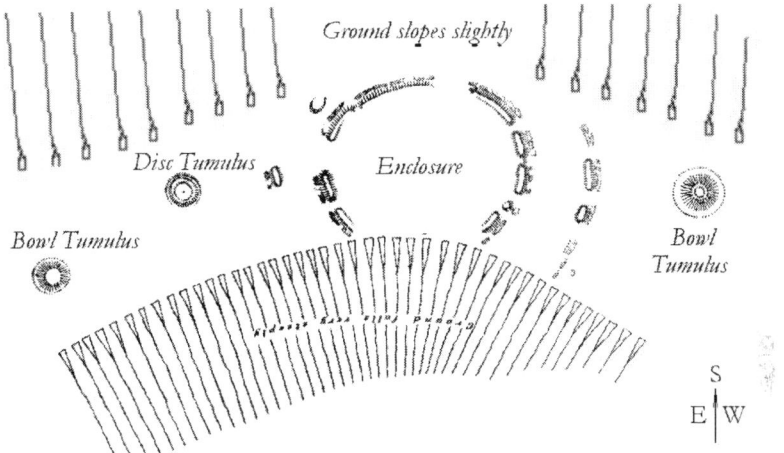

Adapted from Map by Curwen

This layout happens to be is similar to the one used earlier to explain how to calculate the size of the Earth. The layout allows someone standing at 'X' to explain how the maths work, using the simplest number system possible (1:2 slope), to people standing in the southern part of the circle:

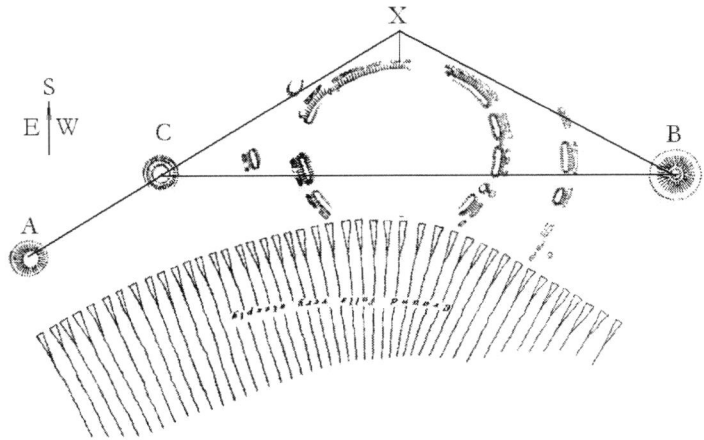

Showing how to calculate the curvature of the Earth

The line from 'X' crosses through a ring tumulus 'C' and sights onto tumulus 'A'. Bowl tumulus 'B' can also be seen from the top rim of the main circle and would be visible to anyone looking from that circle:

Combe Hill: Bowl Tumulus B

All of these elements can be seen from the south side of the main circle at Combe Hill. According to Staveley's resistivity survey,[10] a semicircle of posts once stood at the south of the Combe Hill circle. According to that survey, these are the only posts known to be directly connected with the monument. Anyone standing within the half ring of posts would look directly over the 'X' location and would also see Bourne Hill directly in front of them:

Bourne Hill seen from the semicircle at Combe

This combination; a part circle on a hill, which looks southwards towards places at which the size of the world can be measured, will reoccur elsewhere and is described in later chapters.

Summary

The area around Bourne Hill appears to be exceptionally well laid out to prove that the Earth is curved and to find its size.

The tumuli leading from Beachy Head to Bourne Hill have the appearance of being laid out as sighting platforms. This would allow the height, at any point along the route, to be measured. There are dozens of Neolithic man-made platforms along this route.

Many of these also appear to be arranged with a large platform (as with the Bourne Hill experiment). If this area's purpose were to teach, the layout and number of suitable monuments indicates that this was done on a vast scale.

The area around Combe Hill has the appearance of being laid out to show how to calculate the size of the world. From test demonstrations, at this location on the South Downs, the method is simple to demonstrate at any time of day; and this would allow a very large number of people to see how it works with just the one layout.

The arrangements shown in this part do not prove that the people of the Neolithic knew about the heavens nor the size of the world. However, the monuments appear to be arranged to allow large scale teaching of how to prove what the world is, how to find its size and how to understand the heavens.

It is not possible to say what these monuments were really for. Some may have been constructed later than Stonehenge; perhaps improvements on earlier versions. Nevertheless, the experiments shown in this chapter demonstrate that Neolithic people, especially on the island we now know as Britain, would have had access to materials good enough to find out about their Universe.

8: NAYSAYERS TO A ROUND WORLD

At a place such as Bourne Hill, it is possible to find the size of a round world using a raised flat-top mound, which looks both east and west, by measuring the difference between the 'up' and the 'down' slopes. However, some might argue that the world is a disc, some sort of squashed hemisphere or, perhaps, something else entirely.

One way to check whether or not the world is a something other than a ball is to measure its size from various different hills, each with a different height. In the imaginary picture below, the experiment is undertaken on a squashed hemisphere rather than a ball. For this experiment, and if the world is not a sphere, the radius calculated would be different for hills of a different height:

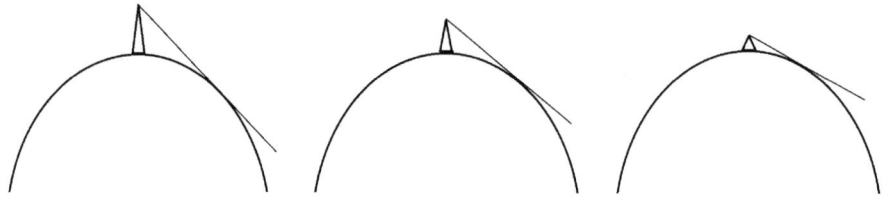

Repeating the experiment at different heights

If the measurements give a different answer when checked at hills of a different height, the world is unlikely to be a spherical ball. However, if the calculation of the measured size at each location is always the same, the world is likely to be spherical; at least in the area near to the hills being measured.

At the far end of the South Downs, some of the other hills, with good views east and west, also have a large flat-topped bowl tumulus at the summit. A few of these are described in the next part of this chapter.

8: NAYSAYERS TO A ROUND WORLD

The rotation of the Heavens

The path from Combe Hill leads down through an ancient field system to a village called Jevington, where it re-joins the path of the South Downs Way coming from Bourne Hill. It leads upwards to the next ridge over, also along the South Downs Way, where several more bowl tumuli exist at the brow of Folkington Hill. These tumuli also look eastwards and westwards over sea views.

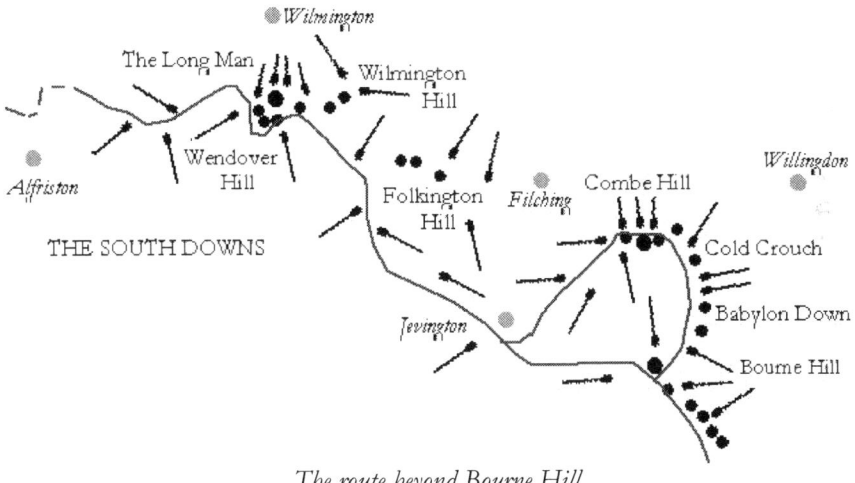

The route beyond Bourne Hill

At a height almost identical to Bourne Hill, Folkington Hill[01] also has a large bowl tumulus at its summit:

Folkington Hill barrow

At the peak of the hill above Folkington, the remains can be found of another bowl tumulus on the summit of Wilmington hill[02] (214 metres above sea level):

Wilmington Hill barrow

This mound also has sea views to the east and west. By walking from Beachy Head to Bourne Hill and then crossing the valley to the next ridge, on which Wilmington Hill is located, more bowl tumuli with sea views east and west can be found. All of these are situated on ridges within sight of other, often smaller, barrows along the same ridgeways.

In each area, large bowl tumuli appear to exist at each hill summit, with smaller barrows leading down towards the sea along the hill ridges. In the view below, from Holt Brow barrow, the remains of four intermediate barrows, leading up to Wilmington Hill, can be found:[03]

The view to Wilmington Hill from Holt Brow barrow

The next hill ridge to the west of Wilmington Hill, above the village of Alfriston, also contains similar arrangements which lead up towards Firle Beacon and its flat-topped barrow.

8: NAYSAYERS TO A ROUND WORLD

The Long Man of Wilmington

Wilmington Hill is just above the Long Man of Wilmington. A few hundred yards immediately to the west is yet another bowl tumulus[04] at Windover Hill. This barrow does not have particularly good sea views and its view to the east is obscured by Wilmington Hill.

Below this barrow is a steep north-facing slope which points upwards in just the right direction to find the polar axis. Here you can lie comfortably on the ground at night and, using two sticks, follow the rotation of the stars. Perhaps by coincidence, another nearby steep north facing slope, at which this can easily be done, is just below the Combe Hill Camp.

This hill was cleared of trees in the Neolithic period. On the steep north-facing slope of the hill is a figure holding two sticks:

The Long Man, East Sussex

In the 16th century, the figure was lined using bricks. It is not clear who decided to construct it, and the reasons for its construction are lost.

The tumulus at the top of Windover Hill[05] looks down over what is thought to be an old quarry,[06] with a north-facing entry to its bowl:

The workings above the Long Man (bowl tumulus seen at top right)

Though its age is unknown, this quarry's north side works well as a late evening amphitheatre; and this duplicates the effect of lying on the Long Man slope itself (which is not accessible). The steep north-facing sides allow comfortable viewing and demonstration of the rotation of the heavens. By using two sticks placed to point to the stars, over a long summer's night, the stars can be seen to rotate in almost a full circle:

Following the rotation of the Heavens on a northern slope

The most obvious explanation for that rotation of the heavens, if the Earth is fixed, is that the Universe is also a sphere.

Two sticks and a hill's height

But two sticks can also be used for a less obvious purpose. If a groove is cut down the middle of a stick, but leaving two small, and equally deep, 'gates' either end; then the stick can be used as a builder's level when filled with water. When water is poured into the middle of the trough and is gradually lifted at one end, (so that the water drips out at the same rate at either end), the hollowed stick is level:

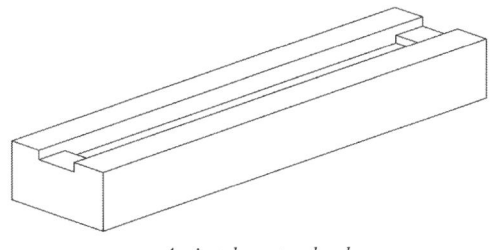

A simple water level

This is effectively the same as a standard builder's level, but a lot more difficult to use. It can be checked for accuracy in exactly the same way as a standard level: if the level is rotated around by 180° and still has water flowing out either end at the same rate, then it is truly level.

A longer stick can be used to get an even more accurate level. Providing the depth of the gate at each end of the stick is the same, the stick does not need a perfectly flat top surface for it to work as a level:

An extended water level

This type of water level can also be wrapped with a piece of hair:

Wrapping hair to make a surveyor's level

If a second stick is marked with bands, the two sticks can also be used as a surveyor's level and staff:

Finding levels

Providing the viewer's eyesight is good, this would allow the height of a hill to be found with some accuracy. In a method similar to the one used today, small calibration errors can be reduced simply by viewing from one end of the stick and then doing the next step by viewing from the other end of the stick. By switching the level around for each leg of the survey, small cumulative errors will largely disappear:

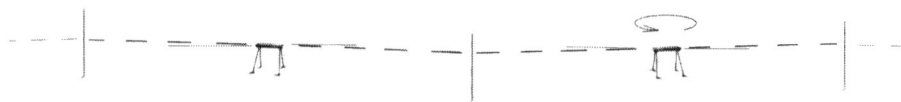

Removing cumulative errors

8: NAYSAYERS TO A ROUND WORLD 105

Two sticks at Wilmington

The same two sticks can also be used to create a level surface to measure the angle down to the horizon:

Creating a levelling platform

Depending on the accuracy needed, the second stick could be relocated onto a third level platform, a forth and so on. This can create a level surface between blocks over a large distance. By turning the level by 180° each time this is done, minor errors in the level can be taken out.

If set above local shrubs on a nearly level flat platform, the procedure can quickly measure the angle down to any horizon. Pieces of wood, counters or a measuring tape could be used to measure the height (at which the other end of the stick is seen to meet the horizon):

Measuring the angle down to the horizon using a timber block

Once that height at one end is measured, it can be used to count along the sticks to find out what the slope down to the horizon is:

Calculating the slope down to the horizon (Folkington Hill)

The number of times (N) that the measured height can be counted along the length between the two ends of the is the slope down (1/N). The value 'N' can then be used in the equation to find the radius of the world: R = 2.h.N.N (where 'h' is the height of the hill above sea level: see "Calculating the size of the world" in the previous chapter.)

Because cloud or mist might obscure the real horizon, the measurement is most accurate when pointed to the horizon at the moment of sunrise or sunset. If done only by eye, there are quite a few ways to measure this height. In the photograph below, coins were used during a springtime sunset:

Using coins: Beachy Head at sunset

At the start of this chapter, it was shown that Naysayers could claim that different heights of hills would give different answers. So to show that the world is a constant curvature, level mounds would need to be built on other hills. Because it only needs a single sight-line down to the sea, this method allows the Earth's curvature to be confirmed at many hills.

This type of experiment could also be done at Bourne Hill (+202m above sea level), Wilmington Hill (+214m) and Beachy Head (about +164m). Earlier in this chapter, and in the previous chapter, flat-topped neolithic mounds were shown to be located at these positions.

In the immediate vicinity, the experiment can also be done at Folkington Hill (+202m), where, together with some smaller mounds, a larger flat-topped mound [01] also exists:

Folkington Hill brow Neolithic mound

And also Willingdon Hill (+184m), where a larger flat-topped mound [07] also exists, together with some smaller intermediate mounds leading towards Bourne Hill:

Willingdon Hill brow Neolithic mound

Together with Fore Down Hill[08] (+134m):

Fore Down hill Neolithic mound

And also Beachy Brow[09] (+168m), Pashley Hill[10] (+171m) and various other locations on the Eastbourne and 'Long Man' ridge. This type of Neolithic mound also exist at hilltops and brows along the inland ridge, west of Alfriston, leading up to Firle Beacon[11] (+217m):

Firle Beacon Neolithic mound

At each of these locations, good to measure and confirm the size and shape of the world, ancient larger flat-topped mounds can be found.

Sunrise and sunset

The location of sunset is easy to predict, however, sunrise is not so easy to predict; especially if setting up in the twilight hours. This difficulty can be removed by having one stick with a very flat surface and then turning it by 90 degrees. Using this method, the measurement can be taken wherever the sun rises:

Looking east to the sea from Wilmington Hill

Looking west to the sea from Wilmington Hill

It's just here. Nowhere else

Another counter-argument against a round world is that perhaps only one location (for instance the South Downs) has the qualities of a sphere. This argument would need more work to be proven untrue but, if measurements at places far away were taken, and the same values were found, the world is spherical.

To find out one way or the other, other locations along the South Coast would need to be checked:

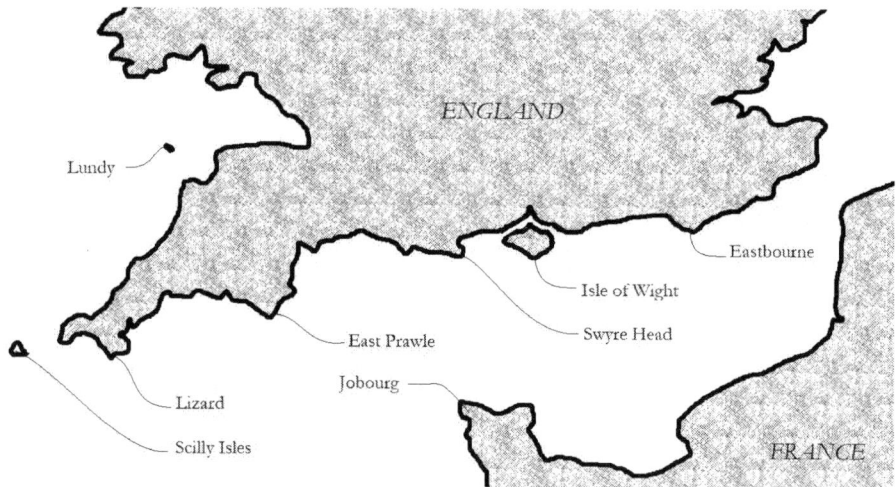

Location of hills with east and west views over sea

At each of the places above; those hilltop locations best suited to do this experiment; there exists either an ancient mound or a Pastscape record of an ancient mound.[12]

Swyre Head, Isle of Purbeck, Dorset

8: NAYSAYERS TO A ROUND WORLD

Summary

It is possible that simple levelling sticks were being used to measure the size of the world. If that were the case, knowledge of channels and how water acts would have been common.

This chapter has shown that some types of bowl barrow, or flat-topped mounds, are located in those locations that would be expected if the world had been measured, possibly using levelling sticks, in these islands. It showed that this is relatively simple to do (using shaped sticks) and that the mounds required also exist at places with sea views in only one direction. The Long Man of Wilmington, though constructed much later, may be the remnants of a folk memory from those times.

The mounds of Folkington (foreground) and Bourne Hills (high background) seen from the mound of Wilmington Hill

It may not be a coincidence that Stonehenge's outer lintels are perfectly level. Some sort of levelling device must have been used to construct the monument.

Unusually, the two hills of Bourne and Folkington (above) are at almost identical heights (approx 202m above ordnance datum to the top of each mound). The shot above shows how difficult it is to work out which hill is the highest, or even that they are level.

Two hill summits, near to each other and of almost exactly the same height, are a rarity. But this rare occurrence also happens at two other locations; both of which are known to be connected to Stonehenge.

9: BEGINNINGS

Before anyone might have decided to try to find out the size of the world, someone else would have needed to notice that the world can not be a disc.

In the mountains of Preseli, there are two peaks, which happen to be at almost exactly the same height, are within easy walking distance of each other, and which happen to have steep sides. An ancient pathway runs between these peaks before descending through a rocky outcrop known as Carn Goedog. Those two peaks are named Foel Feddau and Foel Eryr:

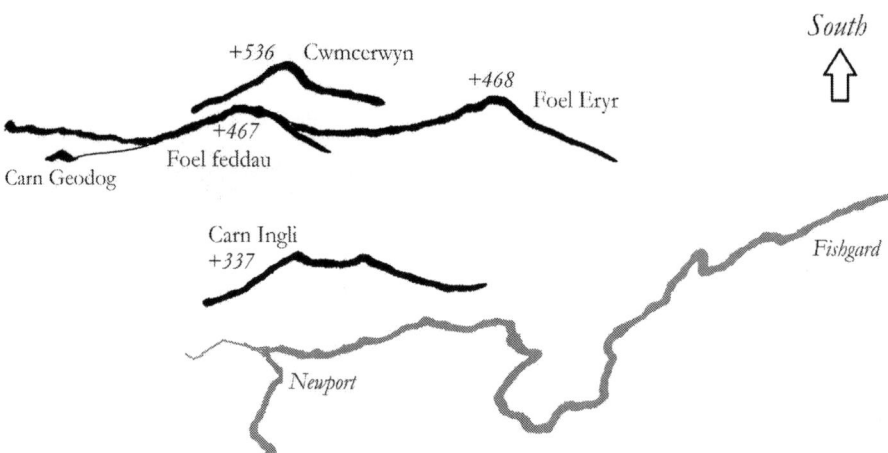

A schematic drawing of the Preseli Mountains

This unusual topographical combination gifts the observer with an unusual visual effect which is similar to that which can be seen at the two (equal height) hills of Bourne and Folkington. However, at Eryr and Feddau, this effect is very much easier to see.

9: BEGINNINGS

Starting at Foel Eryr, Foel Feddau can easily be seen below the slopes of Cwmcerwyn. The arrow below shows Foel Feddau poking just above the horizon:

Foel Feddau seen from Foel Eryr

Walking to Foel Feddau, it is possible to keep easy track of Foel Eryr and, on arrival at Feddau and then looking backwards, Foel Eryr is also obviously visible above the horizon:

Foel Eryr seen from Foel Feddau

This effect occurs because both hills happen, by chance, to be the same height and also to have unobstructed views for great distances. Due to the curvature of the Earth, the higher mountains of the Brecon Beacons, to the far east of Foel Feddau, are so far away that they do not jut up above the horizontal.

These views show that we are not on an endless flat world. For it to be possible to see both hills from each other, and for both to be above the horizon, the world must either be curved or be some form of plate.

As with the hills of the South Downs, there also exists a hill that is set to the North. This is known as Waun Mawn:

Waun Mawn

Recent archaeological reports[01] appear to suggest that, as with Combe Hill, there once existed a partial circular ring at this location. Standing at the southern-most point of that ring of stones, the two hills of Eryr and Feddau can be seen:

Eryr and Feddau seen from Waun Mawn

From this vantage point (Waun Mawn), and having seen each hill from the other, it is possible to explain why the world can not be endlessly flat.

Eryr and Feddau

At each of the two hills, cairns were built long ago. The cairn at Eryr[02] is so large that it can be seen by eye from Foel Feddau:

Foel Eryr's cairn

Foel Feddau, which is probably an existing natural mound, has a more modest collection of stones at its summit:[03]

Foel Feddau and its cairn

It would be relatively easy to find the distance between Feddau and Eryr, estimate the height of Foel Eryr, and to also estimate the height between the top of Eryr and the location at which its northern slope meets the horizon. By doing this, the slope downwards would provide a fast estimate of the distance to the edge of the world (if it were a disk):

The world if a disc

But if this were done, the figures would show that the disc is too small for the land beyond the Brecon Beacons, the land we now call England, to exist. So the Earth must be curved. It is not possible to be absolutely sure, without a detailed survey, that the two peaks are at exactly the same height; so any early estimate of this type could only be approximate.

The Preseli hills are windy and frequently battered by gales from the Atlantic. If a method using levelling sticks were subsequently used at Preseli, a long trough made of timber would be unlikely to stay in place in those fierce winds. On the other hand, if troughs were brought up every time needed, they would need to be lightweight. If they were lightweight, they would be affected by the lighter gusts of wind which are usual for this area. These conditions reduce the options available.

So the choice for a designer would be to either make the trough out of something like stone or to wait it out for a completely still day; a rarity at the top of these mountains.

If made of stone, the trough would need to be solid enough to not vibrate in the lighter winds or to be blown away in the heavy gales. However, even on a normal day in Preseli, the surface of the water would be likely to be blown down the trough.

One way around this problem is to put a timber cover onto the trough and to leave the ends open so that clay plugs can be inserted at either end. Water can then be poured in through a hole in the cover, and the top of the clay plugs gradually lowered until the water is just brimming over at either end. With the cover removed, a perfect level for sighting to the horizon exists between the tops of the clay plugs.

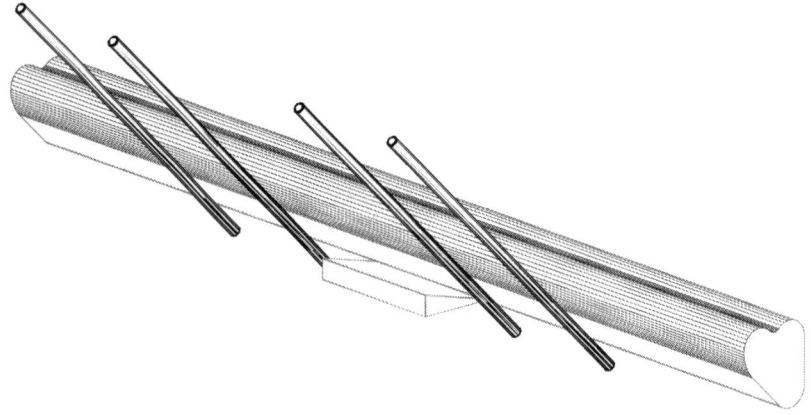

Rotating a stone trough to position

Cwmcerwyn

As found elsewhere, the ideal location for this sort of experiment would be the highest mountain, or hill, set within a promontory which juts out into the sea. In the Preseli area, the ideal mountain to do this, known as Cwmcerwyn, exists just above Foel Eryr and Foel Feddau.

The peak of Cwmcerwyn is 536 metres above sea level:

Foel Cwmcerwyn

Half way up from Feddau to Cwmcerwyn, a small intermediate tumulus exists [04] and, on top of Cwmcerwyn, there is a large flat-topped mound [05] together with an Ordnance Survey marker:

The mound atop Cwmcerwyn

This large mound is accompanied by a small cairn,[06] and a smaller secondary mound.[07] As may be the case for Foel Feddau, the cairn may be the result of treasure hunting and excavation:

The small cairn of Cwmcerwyn

From the top of Cwmcerwyn, there is a massive panorama of views towards the sea. This location is therefore ideal to measure the size of a spherical world using a 'stone trough' method.

The panorama from Cwmcerwyn

But there are no obvious remnants of any such stone at Cwmcerwyn. If this experiment ever happened, the stone trough has been moved somewhere else. However, just below Cwmcerwyn (Carn Goedog; bottom left below) there exists an outcrop of a soft bluestone which is ideally suited to this purpose:

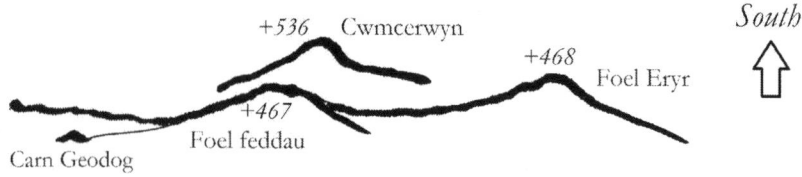

The location of Carn Goedog

This outcrop of stone is easily visible from Foel Feddau:

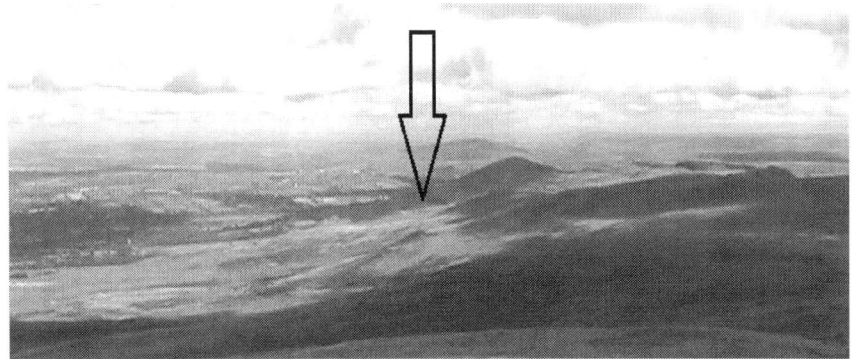

Carn Goedog seen from Feddau

Carn Goedog[08] is known to be the exact source of the bluestone used for some of the innermost stones of Stonehenge.

At Stonehenge, bluestone number 68 is buried deep into the ground. It stands next to the central Altar stone. During the twilight hours, a torchlight can illuminate its deep, and perfectly straight, central trough.

If this stone were brought to Stonehenge, and if, long ago, it had been used to find out the size of the world, then that information must have been so well known that it was no longer felt necessary to keep the tools in place to show how to find it.

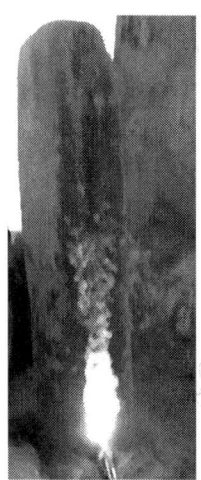

Stone 68

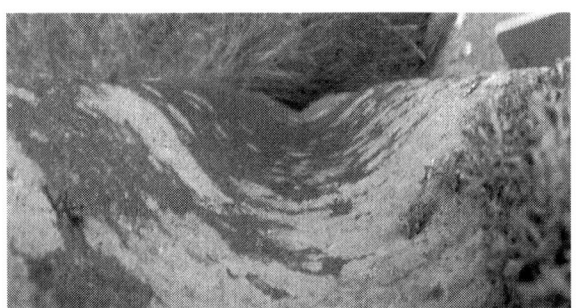

A view down the trough of stone 68

If any of the above happened, the information on the size of the world must, somehow or other, have been encoded into the monument itself.

Numbers and bases

Many people have suggested that measurement units exist at Stonehenge and other monuments. Alexander Thom thought that the "Megalithic Yard", a value of about 0.83metres [09] (2.72 statute feet) was in use. The antiquarian Stukeley estimated a value of 0.528 metres.[10] Many other ideas have been proposed and Flinders Petrie, an Egyptologist who pioneered systematic methodology in archaeology, noted that the internal diameter of Stonehenge is nearly equal to one hundred feet (30.5 metres).

The system of counting we used today is based on the Roman system, which uses two hands for each count. That numbering system can use fingers to count up to 5, with a 'V' being one hand and an 'X' two hands.

However, before the number 10 became universal, many other systems were in use. The Yuki used base 8, the Babylonians base 60, the Aztecs 20 and the Maya used 20 but with a sub-base of 5. Duodecimal (12) and various other systems were also in use in times long past.

The base 30 or 60 system is useful for astronomic calculations because it divides relatively easily into the days of a month or year. Degrees of a circle are still based on that type of system (the 360° of a circle).

It is also easy to count up to 30 using two hands:

Counting to 30 using two hands

A 30 count is also useful for doing calculations on the size of the world. For example, if size 8 feet (about 263mm) were used to find the height of the hill at Bourne Hill near Eastbourne, the hill would be 768 'feet' high (768 size 8 feet, or 202 metres) above average sea level.

768, the height of this hill (h), in a 30 count, is a few feet over *30 x 25*.
Two people, two sets of hands, can count this without using paper

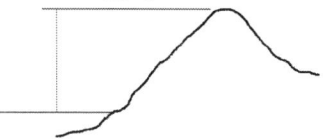

At Bourne, the measured slope to the horizon is about 1:126. This is 6x21 or 30x21/5 using a 30 count system.

The slope denominator (N) using base 30 is: *30 x 21/5*
Two people, two sets of hands, can count this without using paper

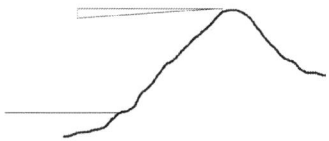

In an earlier chapter, subtitle "The size of the world", the approximate radius of the world was found using the equation R = 2.h.N.N.

At Bourne Hill, and using the numbers above, that equation is:
2 x (30x25) x (30x21/5) x (30x21/5)
Which is the same as:
30x30x30 x 2x25 x 21x21/25
Or:
30x30x30 x (2x21x21)
And *2x21x21* is 882, a little under 900, is roughly 30x30. So the radius of the world, measured in "actual feet" works out as:
30 x 30 x 30 x 30 x 30

A complex piece of maths which ends up with thirty to the fifth power:

$$30^5$$

That figure, (30^5 multiplied by an average human foot), in modern-day measures is:

0.263m x 30^5 ~ 6,391km (3,971 miles). The radius of our world is actually about 6,371 km (or 3,958 miles). So the only numbers that would be needed to remember our world's radius, using only the average foot length,[11] are the numbers thirty and five.

Stonehenge once had a ring of thirty sarsens in its outer circle and five sets of trilithon pairs: These numbers are a design feature of Stonehenge:

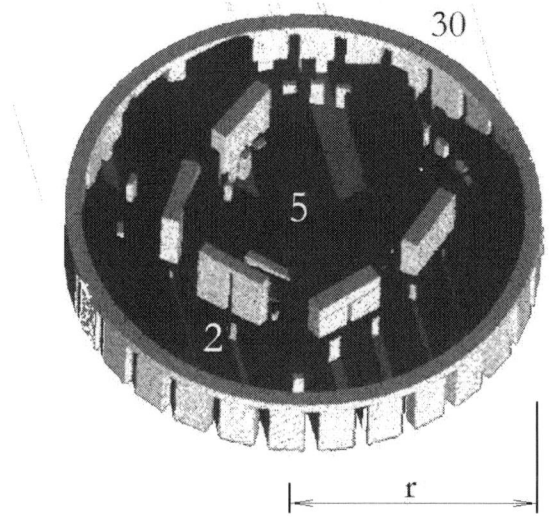

Stonehenge numbers

The trilithon pairs introduce only one other number: The number two.

If the number 2 is multiplied by 30 and then by an average human foot, a value of about 15.78 metres is found. This is the approximate radius (r) of Stonehenge on its *outside* face as seen in the image above (the outside *diameter* is roughly 31.5 metres). Stonehenge's outer face is roughly shaped stone; ideal to represent the hills and mountains of our world.

In other words, Stonehenge's design could describe the size of our planet to scale using a major base of 30 together with the actual average length of the human foot (about a UK size 7 to 9 shoe size). To describe what the Earth's radius is, take one average foot, multiply by 30, then 30, then 30, then 30 and then, 30 again.

Milk Hill, Tan Hill, Foel Feddau and Foel Eryr

In Wiltshire, two other hills happen to be: a) at a higher level than the surrounding countryside; b) at the same level as each other and; c) face each other along a sunset/sunrise axis. These two hills are the highest in Wiltshire and are set within the Mid Wilts Way:

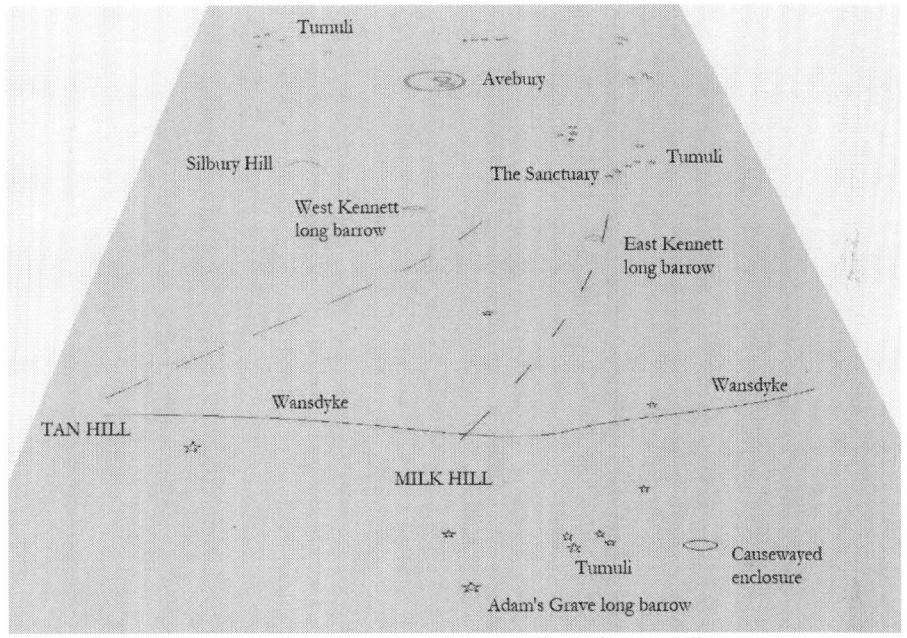

A perspective on the location of Milk and Tan Hills

At 294 metres, the two hills are about two and a half kilometres apart, have a number of bowl barrows,[12] and are connected by the Wansdyke; an ancient vast dyke which extends through this landscape:

Tan Hill and the Wansdyke

As at Preseli, an ancient path (the Wansdyke) leads from one hill to the other. From Milk Hill, Tan Hill is easily seen to stand above the horizon. In the photograph below, the path of the Wansdyke can also be seen winding it's way to the right; where it meets with the horizon:

Tan Hill seen from Milk Hill

Similarly, though slightly less obviously, Milk Hill also stands just above the horizon when seen from Tan Hill. As with Tan Hill, the Wansdyke can be seen winding its way to the left of the hill (in picture below); just where the hill's northern slope meets the horizon:

Milk Hill seen from Tan Hill

From each of these hills, a ridge leads north-east towards a place known as 'The Sanctuary'. On the way, the ridges pass a large number of ancient monuments, including East and West Kennett long barrows. After a short walk (about 6 kilometres or 4 miles), the vast ancient complex of Avebury is reached; just after passing a giant ancient marker known as Silbury Hill (the largest human-made hill in Europe).

Whatever happened here, it was important.

Summary

The area in Preseli, from which the Stonehenge bluestones came, is connected by an unusual topographical coincidence to the hugely important complex of monuments in the area surrounding Avebury. That coincidence is two hills of equal height with ancient monuments at the summit of each.

As with the Preseli bluestones, brought from Wales, the large sarsen stones of Stonehenge were brought from the Avebury area to Stonehenge.[13,14] In this way, Stonehenge's construction links the Avebury area with the Preseli mountains. Stonehenge also contains stones (the bluestones) that could have been used to describe and measure the world's size. Finally, Stonehenge's architectural layout incorporates a number system that could be used to accurately describe the world's size.

The Preseli mountains have topography and monuments with similar features to the South Downs. At both the South Downs and Preseli, there are places (Combe Hill and Waun Mawn) with circular monuments which look southwards towards monuments that can be used to measure the world's size. But in Wiltshire, north of Tan and Milk hills, no natural hill exists at the ideal viewing location.

Instead, a hill built entirely by human hand appears to exist at the ideal location to describe the 'two hill' effect. Though it is unlikely that it was built solely for this purpose, there is the possibility that one of its uses may have been to describe our world:

Silbury Hill seen from West Kennett long barrow

10: FEAR (EARLY MONUMENTS)

Past fear of impending periods of extreme winter is well known and is thought to be recorded on early runestones.[01] Norse legend has it that Fimbulwinter, a winter lasting three years, would be the precursor to Ragnarok: The End Time.

This chapter looks at the remains of structures that might be left behind if a fear of climate change, that the Sun may start to move away from the North, were the motivation to start building. These structures would be designed to find out what is happening rather than try to stop, or ward off, cataclysmic events.

Newgrange: The Sentinel

In this section, Newgrange's layout, internal arrangements and symbols are shown to be the same as those that would be required if people needed to establish whether or not the Sun's spiral 'orbit' is changing from year to year. Its inner stone monument is shown to be capable of being used to focus sunlight, *using a simple method not relying on glass*, to allow extremely accurate measurement. Other solstice aligned monuments, also described in this chapter, appear to have these features.

Newgrange: A blacked out chamber

10: FEAR (EARLY MONUMENTS)

In a world with little pre-existing knowledge, and no modern conveniences, it could have been important to find out whether the place in which people lived would remain safe in the future. In the past, getting anywhere by foot would take a very long time; so a vast effort would be needed if the whole community needed to move.

There are a couple of things that could be taken for granted: the world appears to be fixed and the skies above constantly rotate around the polar axis. But there are seasons in the northern part of the world, and those seasons are linked to how high the Sun rises during the daytime. In winter, the Sun appears to spend its time circling the South and in summer, the Sun seems to circle high over the North. In the North, the Sun appears to rise higher during the summer than in the winter.

Although we now know that we revolve around the Sun, this would not have been obvious to people in the past. On a world where the Earth appears solid and fixed, the Sun might have been thought to have a mind of its own and, possibly, be capable of deciding to spend more time in the South. If you have no idea what the Sun is, anything is possible. If the Sun changed how it moved, the Northern lands would become cold and, from the tales of travellers, it would be well known that even a small change can cause big temperature differences in the lands of the North.

The Sun appears to move from south to north and then back south. It appears to turn on a circle, but also gradually moves each day in what would seem to be a regular spiral. Its movement on that spiral is as if the daily turn of its circle is held on the branch of a tree, swaying each year from south to north, coming to rest in the north (summer solstice), then swaying back to the south before coming to rest again (winter solstice). Two opposing spirals held in place by something else.

The Sun's movement between the solstices would have appeared to be similar to the spiral of a modern metal toy:

Modern metal spiral toy

The fear

A big concern might arise if it were thought that the whole spiral might itself be slowly moving. If the whole swaying branch, yin yang, or whatever it is that keeps the Sun's spiral in place, were to gradually move to the South, the Sun would then spend less time in the North:

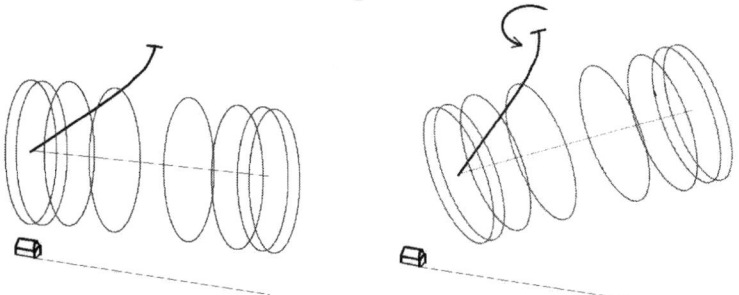

The sway of a swaying branch if the whole tree moves

By the time people knew that the Sun was moving, it could be too late for the community. We know today that this could not happen, but we only know that because we have found that our world revolves around the Sun. This would not have been obvious in the Neolithic: prior to the 16th century, it was widely believed that the Sun revolved around the Earth.

Another body, the one we call the Moon, often appears to go further south than the Sun. The Moon would have appeared to have its own spiral pattern, together with a third cyclical mechanism. This third cycle makes it go to two other extremes that we now call its major and minor standstill (or lunistice). Though nobody would know why the Moon works in this way, it would have seemed to have two major spirals (the same as the Sun) but with a second, much longer, 'waving branch' effect.

Today, we might describe the Moon's progress from its minor to major standstill (over its 19 year period) as following a spiral-like pattern; its apparent movement between standstills due to its orbit following a path skewed at five degrees or so relative to our own orbital path.

However, Neolithic people could not have known why the Moon has this pattern. To them, the Moon's odd path might indicate that the Sun could be capable of doing the same thing, but over a much longer cycle. If it were possible for the Sun to slowly move away on a long cycle, it could be dangerous to remain living too far away from the equator.

Following this way of thinking, one of the first things to find out would be whether or not the 'branch', or whatever might have been thought to hold the Sun's movement in place, is moving. That force might have been imagined to have control over the spiral and, if so, it would have appeared to pull the Sun back to the North over the summer.

The biggest concern would be if the Sun started to go further south each winter. If that could happen, it might be only slight at first. The only time that such a slight change might be measured is winter solstice; the time of its standstill. At that time of year, the location of the Sun's orbit appears to come to a stop and, for just a few days, the Sun seems to circle over the far southern side of our planet without moving in the spiral path.

Over the course of the winter solstice, when the Sun rises and when it falls, a sight line to its exact position on the horizon at sunrise, or sunset, can be found. By recording that position using a couple of sticks, a check could be done each year to see if it has moved southwards:

Measuring solstice using sticks

However, there are difficulties with this method. If it is cloudy, the exact location of sunset (or sunrise) will not be seen and, if the Sun did start to move south, the community would need to know this as soon as possible. But clouds are common in this part of the world and having to wait for a whole year, perhaps several years, for the perfect sunset would be a worrisome solution.

In the Northern Hemisphere, (beyond the Northern Tropic of cancer), the sun always rises and falls at a slant to the horizon. In the Southern Hemisphere, a similar thing happens beyond the tropic of Capricorn. Between the two tropics, and on specific days of the year, the sun rises and falls in a vertical line. But in the far North, those tropics are too far away to be accessible (the nearest latitude, Cancer, is in Africa). So missing the moment of sunset or sunrise during the few days of solstice would lose the opportunity to measure any change in its position:

The line of sunset

So capturing the *angle* of the line of sunrise, without having to see the exact *moment* of sunrise, could be a huge advantage. Changes in the position of the *line of descent or ascent* (Ø) at solstice would also show whether the solstice circle's position is changing from year to year:

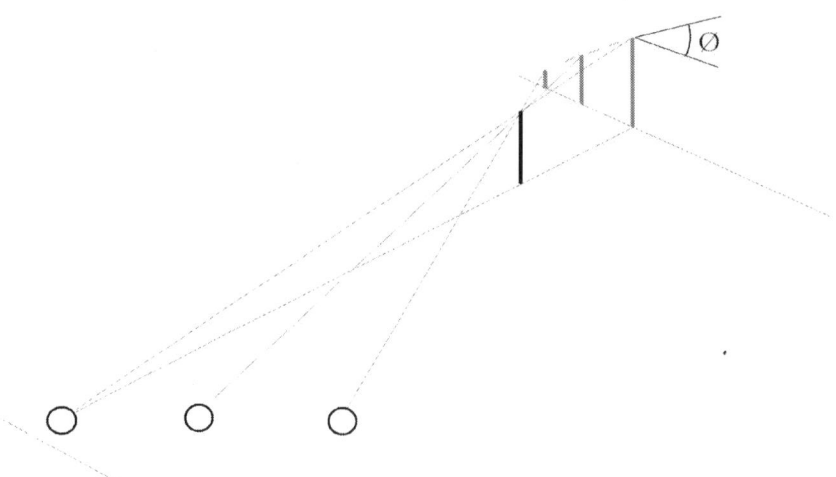

Capturing the angle of sunset using sticks and shadows

The picture above shows a stick being used to cast a shadow-line onto a wall to find the angle of ascent (or descent). One other way to capture the angle of sunset, by using shadows, is to create a long dark corridor or chamber, aligned to the winter solstice sunset (or sunrise). If a very small hole is placed at the entry (to capture light from the Sun), a fuzzy pinpoint of cast light would be seen at the far end of the corridor. At sunset, this will *rise* from left to right (as seen by the viewer).

A trace of that line could then be measured each year. Using this method, the exact moment of winter solstice sunset (at the horizon) is no longer needed: The Sun only needs to peek between clouds for an instant, in the last fifteen minutes or so before sunset/sunrise, to be able to tell if the pinpoint is being cast to the same position on the line.

In the image below, the line at solstice sunset, as seen at the 'backplate' of the chamber, would rise upwards, from the far bottom left to the top right, as sunset approaches. Days either side of the solstice would also have pinpoint lines rising to the right of that line (but these would change slightly from year to year):

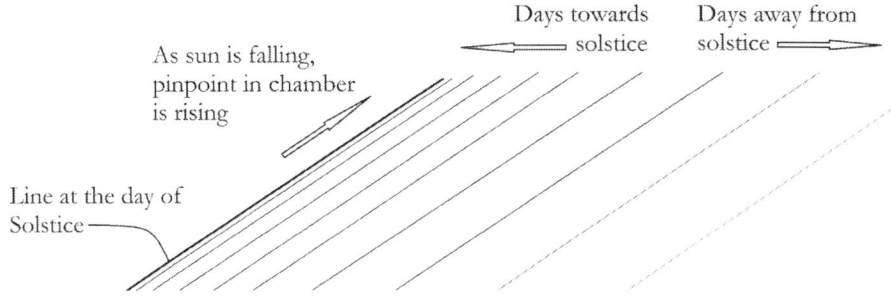

Winter solstice shadow lines at sunset in the Northern Hemisphere

With a sunrise chamber such as Dowth, the pinpoint of cast light would *fall* from left to right. In the image below, the line at winter solstice sunrise falls from the top left to the far right lowest point:

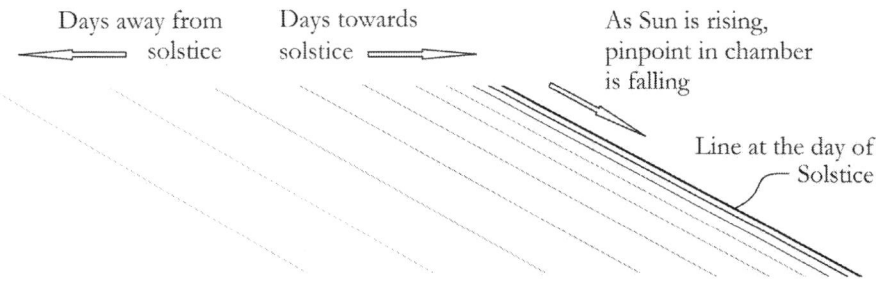

Winter solstice shadow lines at sunrise in the Northern Hemisphere

However, there is yet another difficulty with using cast light to find changes in the Sun's pattern at the solstice. The chamber would need to be long to be able to capture any changes. But because the Sun is roughly half a degree wide, and the chamber long, a dim and fuzzy image of the cast light would be thrown onto the back wall if using a pinpoint source.

In the image below, a shadow from a fence, just three metres away, is much sharper than the shadow from the roof on the right (some twelve or so metres away). So the pinpoint method is good, but it could be difficult to see the precise location at which the 'fuzzy' pinpoint of light falls. For this reason, the other ways to do this experiment (for example using poles set in the ground) might seem almost as good.

Shadows on a wall

But there is a simple method to get over this problem: if a circle shape is inserted into a rectangular entry, it will partially black out the light and form a shadow within the lit area at the back of the chamber:

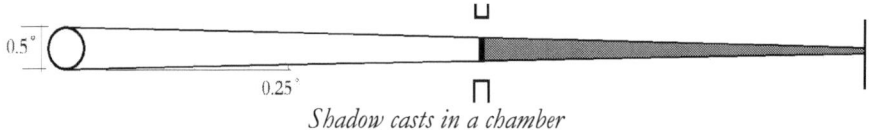
Shadow casts in a chamber

Providing the dimensions of the black-out shape are correct relative to both the diameter of the Sun and the length of the chamber, the shadow formed by this arrangement will still be fuzzy, but easily visible. Most importantly, it will have a completely dark centre point. Bright light falling on a 'backplate', with a central black shadow point, is much easier to see than a pinpoint of dim fuzzy light.

From that 'point of blackness' cast onto the back wall, a more exact trace of the line of the rise/fall of the solstice sun can be found.

A method of testing the fear

Yet another, slightly better, method is to insert a shape other than a circle. The photo below (left) shows the near-shadow of an arrangement when a lozenge, rather than a circle, is fitted into a light-box. Provided the correct focal length is used, the lozenge will cast a faint cross on the far backplate (*the photos below are real and are not computer generated*). At the correct focal point, this shadow, which looks a little like a shamrock when a lozenge shape is inserted, will have a small pinpoint black dot at its centre:

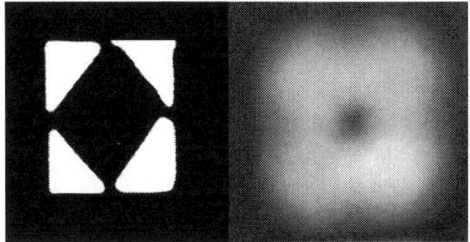

Shadows and effects at a distance

The photographs above show a lozenge shadow next to its cut-out shape (the left photograph). The shadow containing the 'point of blackness', made at a good focal distance, is shown on the right. At other focal lengths, the shadow forms a faint cross rather than a dot.

At Newgrange, a blacked-out chamber was created beneath a mound. During the period in which it was constructed, this chamber was aligned to Winter Solstice sunrise.[02] Above the dark entryway to the Newgrange chamber, a light-box exists at high level:

The entry to the chamber at Newgrange

The light-box above the dark corridor has its own small 'corridor-chamber' (set above the entry walkway of the main chamber). On the lintel above it there are crosses set in squares. It appears that there were originally about nine crosses (8 are shown in the current re-build). This entry box is significantly larger (approx 1.2m wide by 1.0m deep) than the beam of light which penetrates the chamber.

A second shallow internal slot, just beyond the wide entry box, allows a ray of light, some 250mm or so high, to enter the top chamber and fall onto a backplate at the far end. This slot height is the same as would be needed to get a good 'point of blackness' shadow on the backplate. The height of this box, if a 225mm high lozenge were placed within, produces a shadowed black spot some 19 metres or so away (and this is the approximate length of the Newgrange chamber from the internal slot).

Newgrange's light-box

At the rear of the opening at Newgrange is a slot, producing a rectangle of light some 250mm or so deep, above which there is a row of carved lozenges[03] approximately 150mm[04] high.

If a lozenge, or another suitable shape, was to be inserted on the 'shelf' below the carved lozenges (located some 19-20 metres or so from the rear plate), it would produce the 'black dot effect' on the back wall stone of Newgrange. To make that effect, it would be approximately 225 high (165mm along its short faces), and inserted onto the shelf forming the light slot. Alternatively, to produce the 'faint cross' effect, the lozenge would be approximately 150mm high.

The stones of the corridor part of this chamber are likely to have been slightly pushed inwards by passive pressures from the ground. So although the effective slot width (340mm) down the chamber is small today, it may have been slightly wider in the past. Either way, this type of structure would ideally be constructed when it is warm: a shelf that is wider than it is high allows a bit of 'wriggle room' to change things later. (If the location of the slot was built out of position, the shelf arrangement could allow for adjustment and for subsequent improvements).

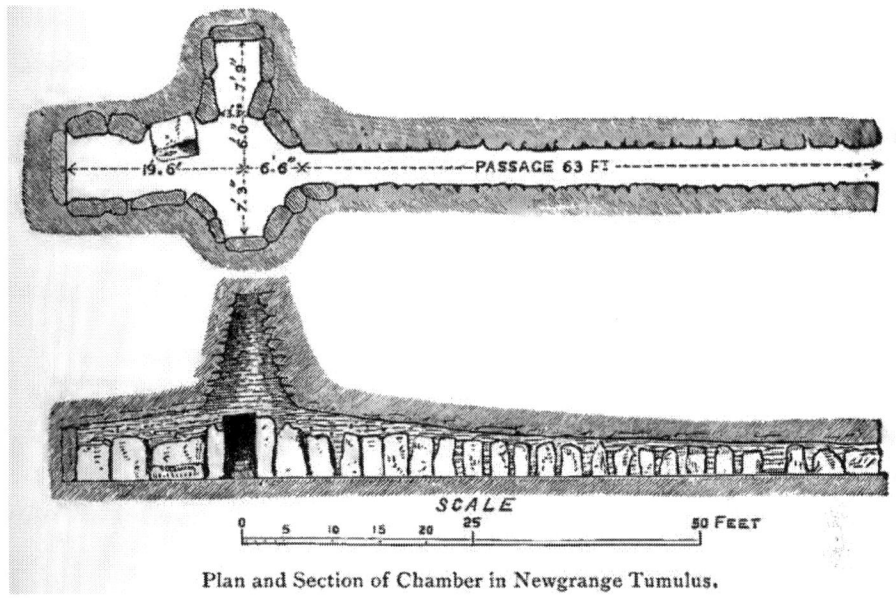

Plan and Section of Chamber in Newgrange Tumulus.

Newgrange: Public domain graphic (extracted from Wikipedia)
File:Wakeman Newgrange tumulus chamber cross section.png (1903) [05]

The 'point of blackness' shadow-line, from various experiments we have done, would be discernibly different, if looking for changes in the line's position, of about an inch (25mm).

At the rear of the Newgrange chamber is a back-plate stone, just like the back-plate (the 'negative' film-roll) of a camera. This plate appears to be in just the correct position to allow the Solstice Sun's shadow's light to fall onto it during the era that it was constructed. The angular change that could be picked up by this arrangement, if the 'point of blackness' method were used at Newgrange to throw light to the back-plate, would be in the region of 25/19,000 (approx 1:750).

To either side of the passage are places where people could have stood to view the solstice event. Above that is a chamber containing enough air for people to cram themselves in, over the ten to twenty minutes or so needed, to find the position of the line of the solstice Sun.

The position of the line of fall of the shadow within the far chamber at Newgrange also seems well suited to allow one person to mark, perhaps with chalk, where it fell.

Newgrange's layout, arrangements and symbols are the same as those required by a system to establish that the Sun has a fixed spiral 'orbit' (one which is not being pulled away to the south). Its inner monument appears to be capable of being used to focus solar light in a dark environment to allow extremely accurate measurement of the location of the solar extreme position during the solstice.

Intertwined spiral symbols at Newgrange

A summary of the sentinels

If finding a way to tell if the lands of the North would become cold were the purpose of Newgrange, then it would have been a sentinel of climate change constructed by exceptionally intelligent people working with limited means. Every year, the sentinel would show that nothing has changed; even if it had been a bad winter and the weather appeared to be saying otherwise.

A similar 'late afternoon' chamber exists at Dowth (near Newgrange). However, this explanation leaves the purpose of the bowls (which exist in the chambers at Newgrange, Knowth and Dowth) unexplained. An explanation for Knowth may help to resolve the unexplained.

The arrangements shown in this chapter do not prove that the people of the Neolithic were concerned that the Sun might move away. However, the monuments seem to be arranged in a way that would allow them to find out.

If laser scanning of the stone at the far end of Newgrange's chamber reveals traces of lines drawn to fall downwards, from left to right and at a shallow angle, then this will have been the purpose of Newgrange: this is all that what would remain if the 'point of blackness' method had been used. Whether traces of such marks would be good enough to survive for 5000 years is not known.

After a few decades, perhaps centuries, it would be known with great certainty that the Sun is not changing its spiral movement (after two centuries, the movement would still be indiscernible).

But if this check were done over millennia, the accuracy at Newgrange is sufficient to find out that the Sun is in fact 'moving' as the tilt of the Earth gradually changes its obliquity over its 41,000 year cycle. The position of the cast light-shadow would move to the left by some 300-400mm; eventually landing on the side wall rather than the back stone and then appearing to rise up the chamber as it hits the floor level.

The effect described above, of light being cast to the floor at the solstice, is what is seen to happen during modern solstices at Newgrange.

Other chambers not aligned to solstice

The Newgrange complex also has a chamber (Knowth) looking in the approximate direction of equinox. If this chamber were used for a similar purpose, but to address a completely different concern, the line of a shadow cast within the chamber would also follow a sloping pattern.

Unfortunately, Knowth is not accessible. However, two similar mound-chambers exist near Avebury, some distance north of Stonehenge. East Kennet, whose chamber may have aligned to solstice, is not accessible. However, West Kennet, aligned to equinox *sunrise*,[06] can be accessed:

West Kennet Long Barrow

As with Newgrange, the main passage has sub-chambers either side. At the far end of the main passage, 'backplate' stones look back down the passage. The slope of the left hand 'backplate' rock approximately mirrors the sunrise trajectory of the shadow-line that would have been cast onto the back face. That piece of sloped stone falls, as would be required, from left to right (see below):

Inside West Kennet Long Barrow: (Stone surface at rear falling from left to right)

Similarly, Loughcrew Cairn T is aligned to equinox *sunrise*:[07]

Loughcrew Cairn T

And contains a similar passage:

The passage into Cairn T

At the rear of this are markers which, if the entrance were set up with a shadow-box, could also be used to follow a cast light shadow at the equinox sunrise:

Loughcrew backplate markers

11: EXPLAINING THE FEAR

If any of the above methods were used to find out how far south the Sun travelled during its winter-time journey, it should be possible to represent those ideas using some sort of drawing. In an earlier chapter, Stonehenge's layout was shown to be the same as a 'fixed world' representation of the Cosmos:

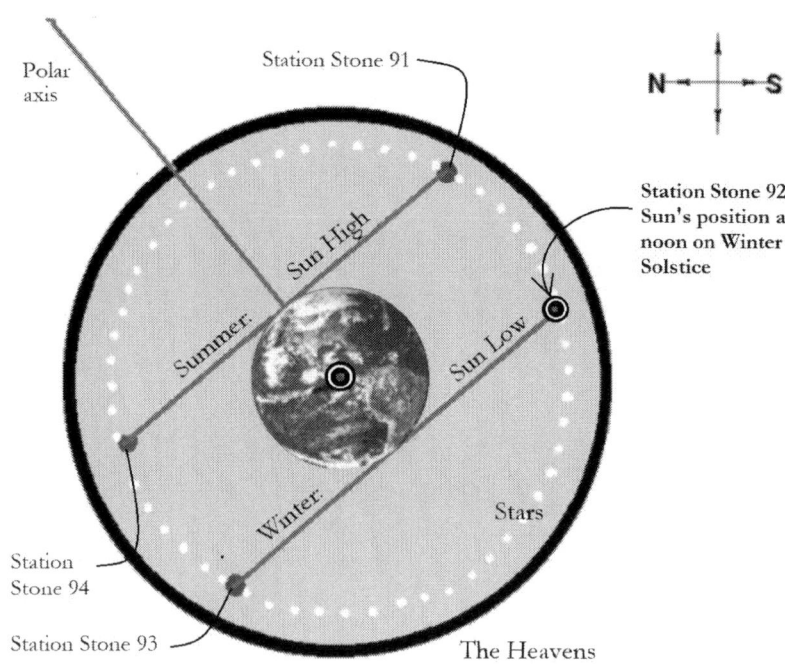

A geocentric heavens: The position of Winter Solstice

In the image above, the Sun's circle of rotation is seen 'side on'. The upper line between midnight (Stonehenge's Station Stone 94) and noon

(Stone 91) shows how the Sun moves at the summer solstice. Similarly, the Sun's winter solstice circle is shown as a line between Station Stone 93 and Stone 92. Today, we might represent this as a graph (for example see CIBSE graph in Appendix A: "Sun-path diagram for latitude 52° N").

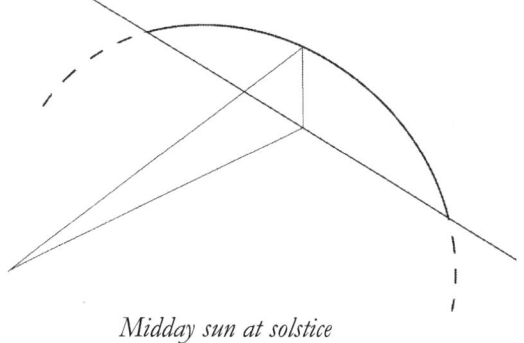

Midday sun at solstice

Explaining the rise and fall of the Sun's circle is difficult using just two dimensions. But one relatively simple way, to explain why the Sun follows the line of a part circle when above the horizon, is to draw a circle representing the Earth and then draw a second circle showing how high the Sun rises above the horizon in Winter. With a horizon marker in the distance, one person can pretend to be the Sun and walk slowly around the circle. This simple method shows how the Sun rises and falls:

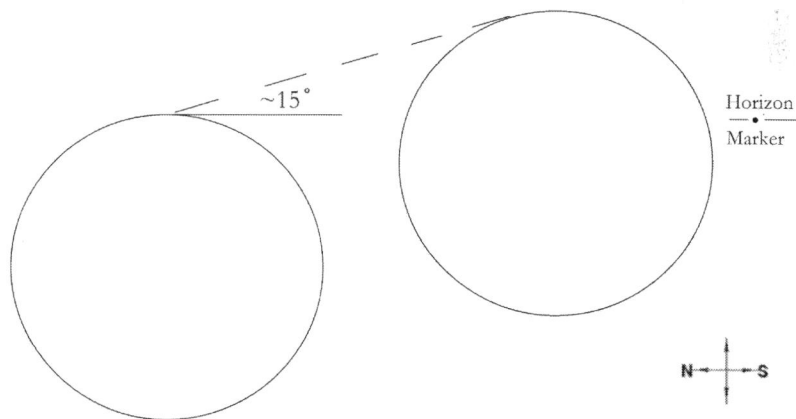

Circles showing the Sun's Circle at Solstice

The arrangement also explains why the Sun rises to such a shallow angle at midday during the period of the winter solstice.

An explanatory layout

If we do not know how far away the Sun is, the position of the Sun at Noon can also be represented as a point from the centre of our World:

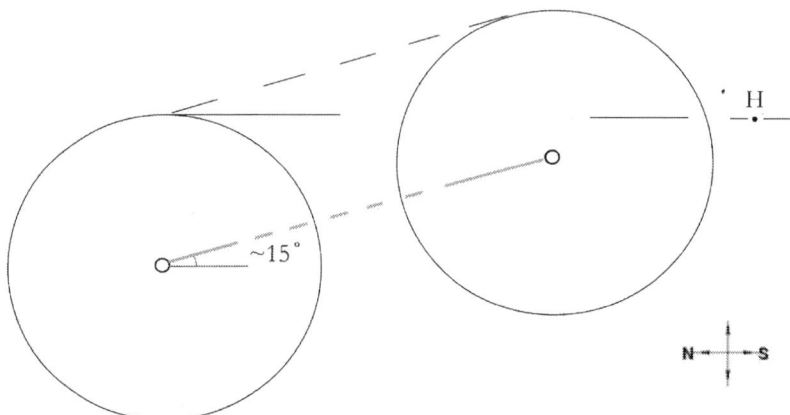

The Sun's angles at Solstice

The Moon, which may have seemed to be going further south than the Sun at its major standstill, might make us worry where the Sun could move to in the future. So the angles that the Moon is measured as rising to, during its minor and major standstills, could be drawn. In the image below, the moon's standstills are shown as lines. If the Sun did start to appear to travel as far south as the Moon, it would only just peek above the horizon during winter:

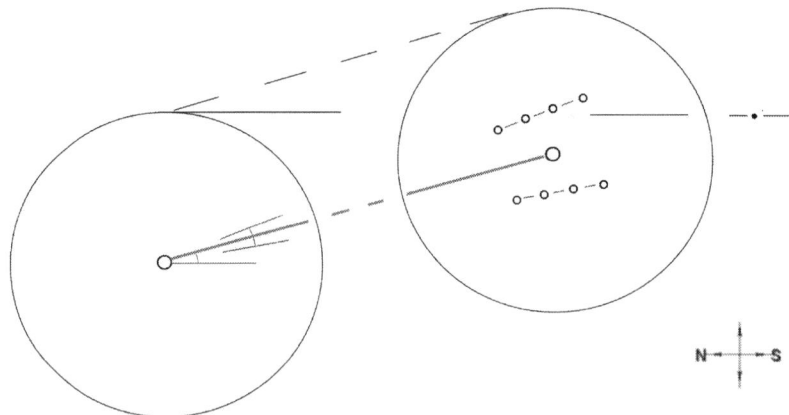

The Sun and Moon: Southern (sun) Solstice with major and minor (moon) lunistices

The stars disappear behind the Moon. So the stars and Universe, in whatever shape they form, would be known to exist beyond the Moon. A surrounding circle is shown below to represent the rest of the Universe:

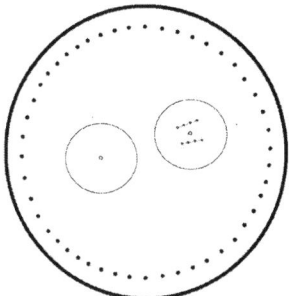

Solstice and Standstills set in a geocentric Universe

The above plan arrangement, with the correct angles to show solar and lunar movement, is the same as found at Avebury in Wiltshire:

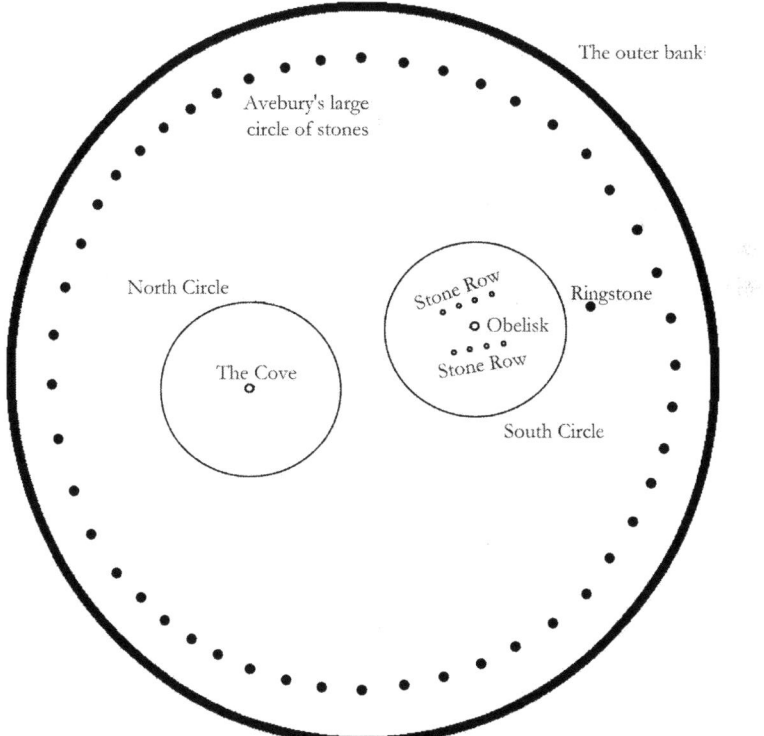

Avebury's layout (with East at top of map)
Indicative layout (not to scale or showing exact number of stones)

The outer bank at Avebury is composed of a massive ditch and bank:

Avebury's bank, ditch and outer stone circle

The circles are made from large stones. The ones at the centre of the North Circle (below) are known as "The Cove". One of these stones has a slope which points approximately to the North Pole:

The Cove at Avebury

Close to Avebury, Milk and Tan Hills are surrounded by large Neolithic earthworks. These two hills could be used to show that the Earth is not flat. When combined with information from elsewhere, especially the hills of the South Downs and of Wales, people could have said with some certainty that the Earth is most likely to be round.

The nearby long barrow at East Kennett could also be repurposed to show that the Sun does not travel further south from year to year. It is possible that the other monuments in the nearby vicinity hold similar, but yet to be discovered, purposes.

Beginnings and endings

In an earlier chapter, a method of checking the position of the Sun's circle was looked at. This is relatively simple and can be done with sticks, or a more complex device such as exists at Newgrange.

The Moon also has lunistices,[01] and these occur once every two weeks or so in a swaying cycle which lasts for nearly a month. However, the lunistices occur at different positions on the horizon. These positions are cyclical over an 18.6 year period and are due to the lunar orbit being tilted at just over 5 degrees to the ecliptic plane. The effect of this is that the measured position of a lunistice varies back and forth, like a swaying branch of a tree: one sway every 9.3 years. To complicate matters, the Moon is also swaying back and forth, like a swaying branch of a tree, over the 28 day period: one sway every 14 day period.

The most northerly (or southerly) position of the *maximum* sway marks a major standstill of the moon. The most northerly (or southerly) position of the *minimum* sway marks a minor standstill:

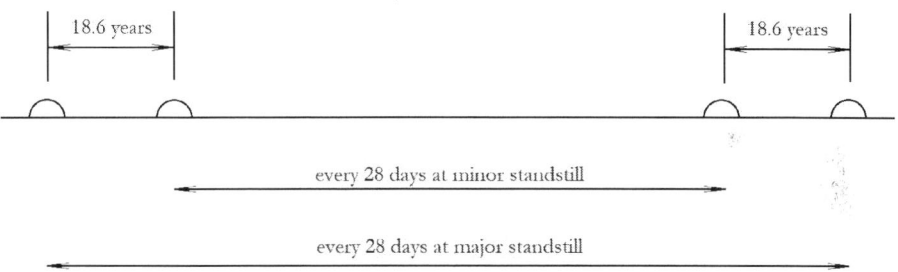

Lunistice variation when seen at a horizon set
Note this is not to scale. Size of moon enlarged for clarity.

From the perspective of a geocentric observer, the Moon is furthest away from the North at its southerly major standstill. If you were to try to find how far the Moon can go, one way to do that would be to use sticks. One stick, in the distance, would be a fixed sighting point and another would be moved to sight through to the distant Moon at the end of the month in its 28-day cycle.

But after a few monthly cycles, the Moon's alignment would be found to have moved. One way to find out what is happening is re-check a 28-

day cycle of the Moon every year or so and, at the maximum point, place a new alignment stick in the ground. After a few years, the maximum alignment would be found:

Posts set to find the major standstill of the Moon

Known as the 'A' posts,[02] a set of post-holes, such as above, does exist in the north-west part of Stonehenge. They are on the correct alignment, and are at about the right positions, relative to the centre of Stonehenge.

However, if the purpose of the original monument were to record the position of the moon and, perhaps later, show the latitudes of the stars, all of that construction was removed with the construction of the new stone monument we now know as "Stonehenge". Similarly, entrances to some of the barrows were purposefully blocked:

The blocking stone at West Kennett

If the order of the Cosmos was written into Stonehenge, the places that might cause fear to return appear to have been removed or blocked off.

Summary

The arrangements shown in this chapter do not prove that the people of the Neolithic were concerned that the Sun might move away or that measuring the Moon's position might have been one of the associated concerns. However, the archaeological findings, and arrangement of existing monuments, seem to be set out in a way that could have allowed the people of that time to find out. After Stonehenge was built, many of the monuments, and particularly those which could have been used to address fears, appear to have been blocked up.

It is not possible to say what these monuments were really for; and the evidence that Avebury was used as a method of recording those fears is a bit weak. Nevertheless, the ruins at those locations are suited to the purpose.

If Stonehenge represents Neolithic knowledge of the Cosmos, other monuments do exist in the location, setting, arrangement and type which would have allowed people the best opportunity to prove that the knowledge was true (although these days we know a little more about how the Universe really works).

The village of Avebury

12: FOLKLORE

If Stonehenge were used to explain the Universe, would there be evidence in folklore of such a dramatic event? It has been shown that folk memories of past events stretch to a time long before Stonehenge. [01]

This chapter looks through some of the other mythological evidence and coincidences.

The Grail

Providing the dimensions are very similar, or identical to, those at Stonehenge, the invention described can produce a very bright 'mini-sun' which might demonstrate a geocentric world's principles. Regardless of the season, the 'mini-sun' appears at about the same height if a seasonal adjustment device is used (the 'three season device').

The castle-like structure of Stonehenge appears to be an ideal structural arrangement for such a task. It is constructed of a very hard material which will not deflect much and can survive the occasional mishap when moving large poles or heavy mirror-laden timber frames.

The process of making a bright 'mini-sun' requires some additional components to be brought in procession to the inside of the round castle-like support structure.

The extra parts required are:
→ A tree-pole or giant lance, pointed to the North Star;
→ A hanger; rather like a candelabra because it carries a light and has a cross-bar which can be rotated;

- A cup-shaped reflector, possibly containing crystals;
- A silvery dish composed of arrays of polished flat metal.

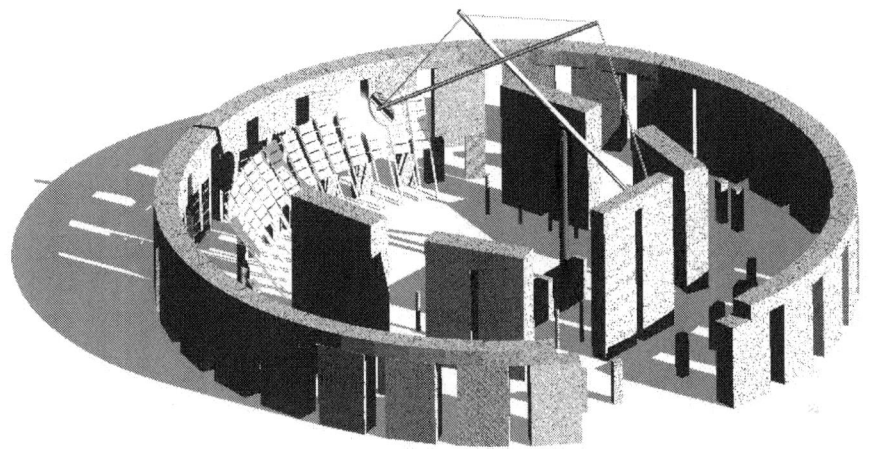

Legend: The possible components of the Grail

In the legend of the Arthurian Grail Procession, the grail is accompanied by several other objects which are brought into the Fisher King's castle. Perceval could have restored the Fisher King, but kept his silence and did not ask the question: "Unto whom one serveth the grail?". When Perceval awoke the next morning, the castle was empty.

At the castle, Perceval sees a strange procession passing by: a squire with a white lance, from which a drop of blood falls on his hand; two squires bearing candelabra; a noble maiden carrying a graal (a receptacle set with precious gems and shedding a brilliant light); and another maiden with a platter of silver.

In the original version,[02] Chrétien did not write about the grail as if it were a religious object; only the later writers call it the "Holy Grail".

> "...Qui une blance lance tint, Enpoingnie par emmi leu; Si passa par entre le feu....Pour çou ne le demanda mie. Atant dui varlet à lui vinrent, Qui candelers en lor mains tinrent De fin or ouvret à chisiel....Un graal entre ses ii- mains Une damoisièle tenoit.... Atout le graal qu'ele tint, Une si grans clartés i vint Que si pierdirent les candoiles Lor clarté, com font les estoiles Quant li solaus lïeye ou la lune.... Pïeres pressieuses avoit....Et li sire au varlet commande.. Qui tint le talléoir d'argent...."

If a reflector is used with the solar device described earlier in this text, the air around the 'great eye' of the reflector becomes hazy due to the concentrated sunlight. The 'candelabra' pole constantly rotates to follow the Sun; so the bright reflector would seem to be covered by haze and would appear to move with sunbeams, fading as they fade.

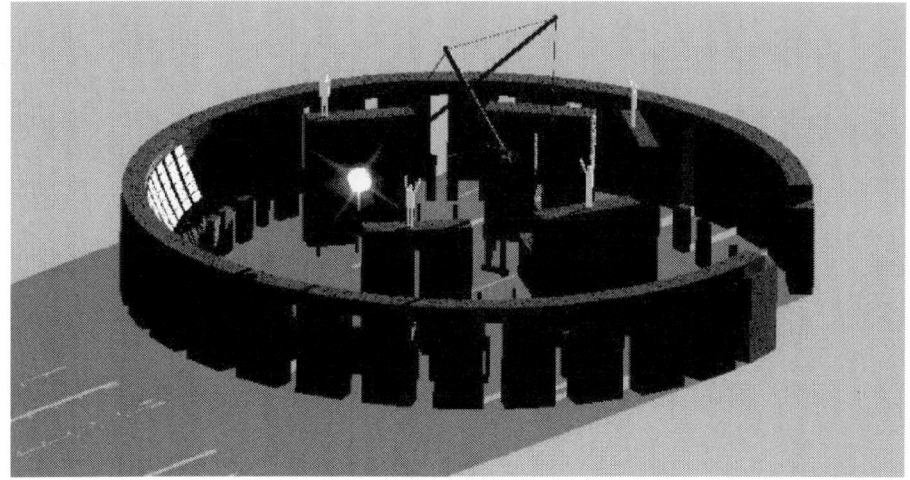

Legend: A shining cup

In the book 'Celtic Myth and Arthurian Romance',[03] the Grail's entry to Camelot is also described as if it were a solar concentrator:

> "It is Pentecost and Arthur and the knights of the round table are at Camelot. Every seat is full and because of this fulfilment Arthur states that *"it is the hour of the Glory of Lorges (England)"*. Even as Arthur speaks there blew a great wind about the castle and a mighty crash of thunder shook the place, then on a sudden a sunbeam cut through the gloom from end to end of the great hall, seven times more clear than ever man saw on the brightest day of summer ... Then the Holy Grail entered into the hall covered in a cloth of white samite, so filled with glorious light that none might behold it. Nor could they see who carried the Holy Grail, for it seemed to glide upon the sunbeam ... then on a sudden it departed from amongst them, and none might see where it went: but the sunbeam faded also."

Curiously, all of the components of the Grail Procession are the same as required to make Stonehenge into a solar concentrator. Chrétien's description of the Grail also happens to fit a description of the effect of this particular type of geocentric solar concentrator.

Treasures of the Tuatha Dé Danann

The Tuatha Dé Danann were a mythical race of Ireland who replaced the people known as the 'Fir Bog'. They were thought to have been in Ireland at about the same time that Stonehenge was in use. They are also thought to have left Ireland at about the same time that Stonehenge fell into disuse (1500 BC).

The process of making a bright 'mini-sun' requires some additional components to be brought into a round castle-like support structure:

- → A tree-pole, or giant lance, pointed to the North Star;
- → A hanger and cup-shaped reflector;
- → A silvery dish composed of arrays of polished flat metal;
- → A stone structure, such as at Stonehenge.

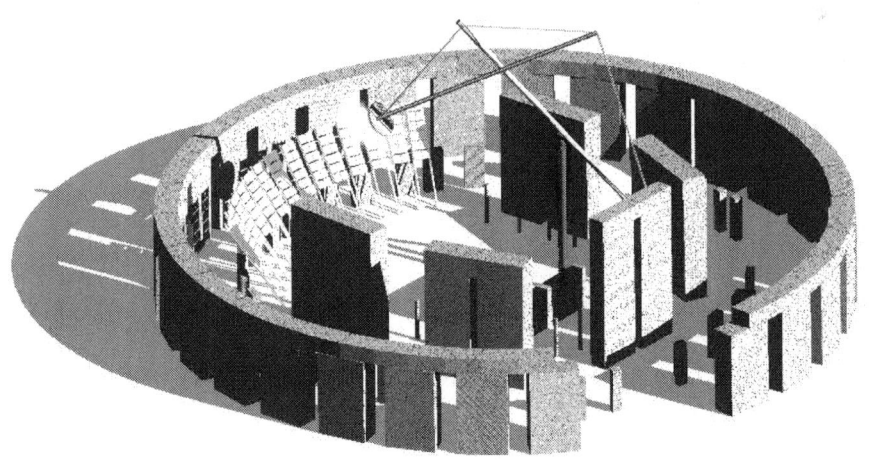

Legend: Tuatha Dé Danann

The Four Treasures of the Tuatha Dé Danann were:

- → The Spear of Lug;
- → The Sword of Light;
- → The Cauldron of the Dagda;
- → The Stone of Fál.

These four treasures could also describe Stonehenge; if it were used to demonstrate a geocentric Universe.

The Druids

The druids of the Isles of Britain have passed down little or nothing of their tradition. Much of what is known of them comes from writings of Greek and Roman scholars. Of what is known, the records seem to agree that Druidry originated in Britain, the primary teaching was reincarnation, and that the Druids claimed to know the size of the world and movements of the heavens. A few samples are:

> "They have, however, their own kind of eloquence, and teachers of wisdom called Druids. They profess to know the size and shape of the world, the movements of the heavens and of the stars, and the will of the gods."
> *Pomponius Mela, "De Situ Orbis", iii, 2, 18 and 19*

> "They likewise discuss and impart to the youth many things respecting the stars and their motion, respecting the extent of the world and of our earth, respecting the nature of things, respecting the power and the majesty of the immortal gods"
> *Cæsar, C. J., "De Bello Gallico", vi, 14 (MIT translation)*

> "This institution {Druidry} is supposed to have been devised in Britain, and to have been brought over from it into Gaul; and now those who desire to gain a more accurate knowledge of that system generally proceed thither for the purpose of studying it."
> *Cæsar, C. J., "De Bello Gallico", vi, 13 (MIT translation0*

"for the belief of Pythagoras prevails among them, that the souls of men are immortal and that after a prescribed number of years they commence upon a new life, the soul entering into another body."
Diodorus Siculus, "Library of History", v, 28:(Loeb/Thayer)

"At the present day, Britannia is still fascinated by magic, and performs its rites with so much ceremony that it almost seems as though it was she who had imparted the cult to the Persians. To such a degree do peoples throughout the whole world, although unlike and quite unknown to one another, agree upon this one point."
Pliny, "Nat. Hist.", XXX, 13: (Thayer)

These descriptions appear to describe a people who had once solved the magic and mystery of the world and the heavens.

Drych Haul Cib Dâr

- → Drych: Mirror, looking-glass; reflection.
- → Haul: Sun, sunlight.
- → Cib: Vessel, bowl, cup; coffer, casket; brim, edge (of vessel).
- → Dâr: Oak-tree.

In a singularly unusual book published in 1914 ('Prehistoric London'[04]), the author (EO Gordon) claims to have been told that "The Druids, it is said, by means of a most powerful reflecting mirror of metal called 'Dyrch Haul Kibddar,' filled the circle [Stonehenge] with a blaze of glory from on high. This is mentioned in the Triads as the *speculum of the all-pervading glance*, or the searcher of mystery; one of 'the Three Secrets of the Isle of Britain.'"

The meaning in Welsh of "Drych-Haul-Cib-Dâr" is "Mirror-Sun-Bowl-Tree". In a peculiarly unlikely coincidence, this list of components also describes the arrangement needed to fill the circle of Stonehenge with "a blaze of glory from on high".

Summary

Stonehenge appears to be laid out as a geocentric description of the world. Its inner monument appears to be arranged, as a geocentric demonstrator, to create a ball of light rising above a round table. And that table appears to be laid out to represent the world.

Stonehenge's plan layout can be shown to be the same as an idealised geocentric description of the Universe. Its inner stone monument is demonstrated to be capable of producing a spectacular public display of solar movement.

The demonstrator's components appear to be the same as those described by Chrétien de Troyes when writing original stories about the Grail Procession. Descriptions of the grail also appear to be the same as a description of the device.

The demonstrator's components also appear to be very similar to the Four Treasures of the Tuatha Dé Danann; a magical people of Ireland's past who, according to likely dates, lived in Ireland at about the same time as Stonehenge was in use.

Descriptions of the Druids, as made by Cæsar, seem to tell of a people who had solved the mystery of the Hinge of the Heavens.

As described in Theories, Chapter 3, Professor Hoyle noted that no account of any previous hypothesis's achievement has been made in *"the full light of documented history"*. The myths described in this chapter do not prove that the selected myths describe the contents of this book. Nevertheless, there are myths that can be shown to have similarities with the events described herein.

13: THE MAKERS' MARK

If the geocentric hypothesis is correct, it might also be possible to deduce where a "makers' mark" would be written into the Stones. If the outer face of the stones of Stonehenge represent our world, the location of England would be at the top of the world within the diagram shown below (the location shown as 'You are here' below):

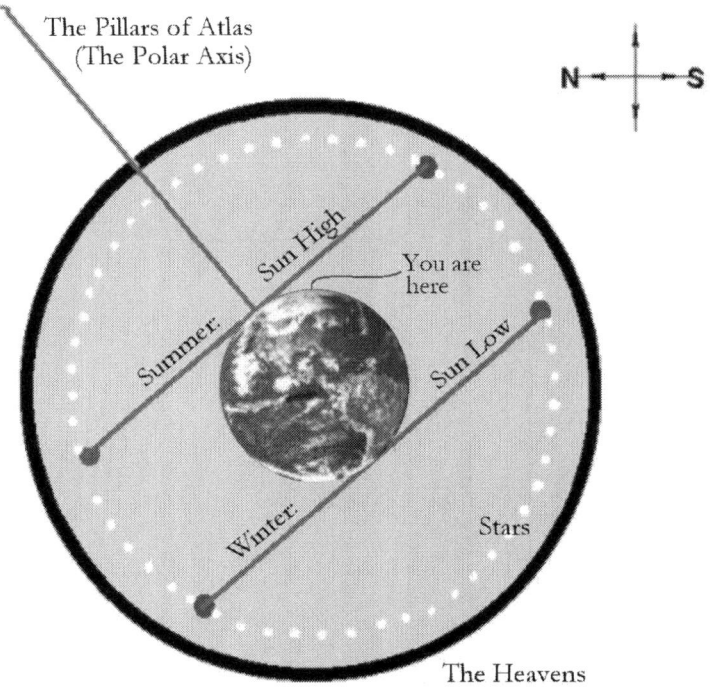

From the previous chapter "Stonehenge and the Hinge"

If the number 30 was special to the people who built Stonehenge (see chapters 5 and 9 for detail), the world described above could be divided into 15 or 30 'latitudes', with each representing a particular circular zone around the world.

In the image below, the 3rd latitude zone is above Iceland, together with the far Scottish Isles & Northern Scotland. The 4th zone is above England, Ireland and the northern edge of France. The 5th zone is above France, Spain and most of continental Europe:

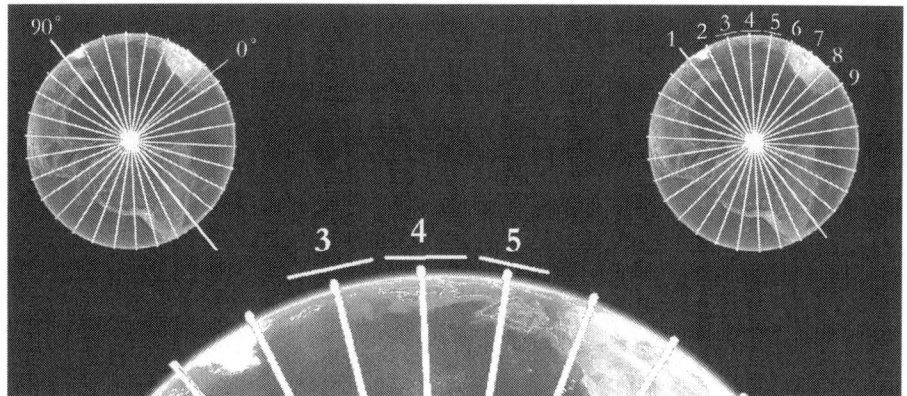

Spokes of the World: Image partly based on Google Earth software

This arrangement of latitudes, using 30 zones (each from a 'spoke' drawn onto the world), neatly divides Northern Europe into three bands, each with sea separating the lands:

3: Scottish islands
4: Britain and Ireland
5: Continental Europe

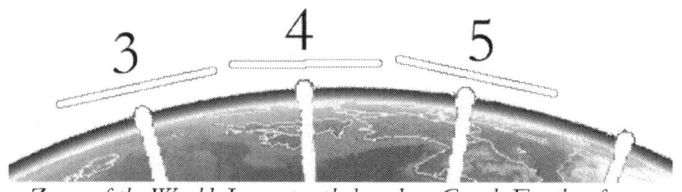

Zones of the World: Image partly based on Google Earth software

If a diagram of that world-view is imposed into a plan of Stonehenge, it is possible to see which stone would have been seen to represent which place. Stone 4 is at the location which would represent Britain and Ireland: due east of the centre. This is at the "top" of the world as seen below; the place you are standing if you believe that you are on a fixed world:

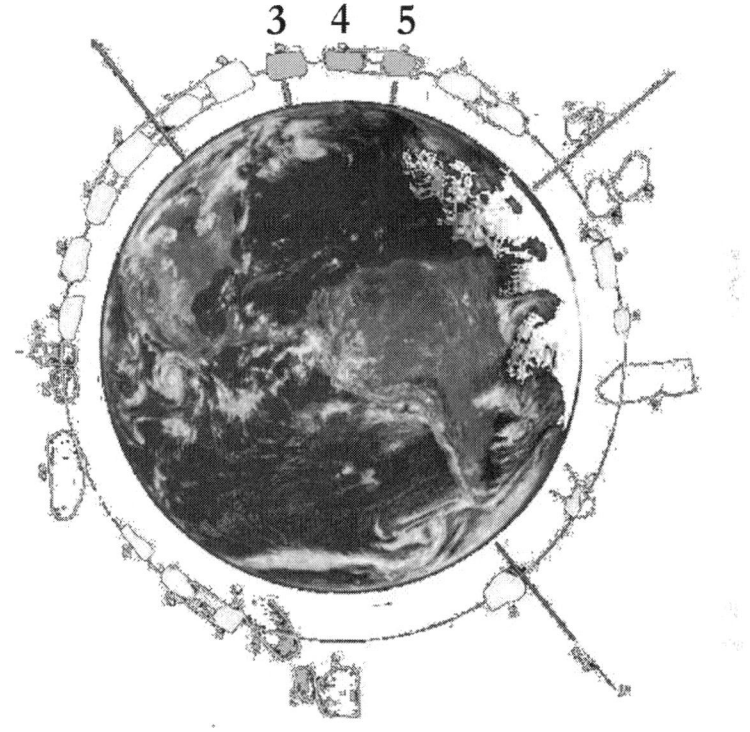

The World imposed on a plan layout of Stonehenge

Using that 'non-moving', or fixed world, thinking, Stone 3 then represents the Scottish islands, and areas as far as Iceland, and Stone 5 represents Continental Europe.

If the makers left a mark to identify themselves, the outside face of those three stones is where the mark, or marks, would be expected to be found. Fortunately, the stones have recently been scanned. The 2012 scanning report shows that 'axe-heads' are all found on the external faces of Stones 3, 4 and 5, together with some on Stone 53 (see Chapter 6).

Stone 3: *Scottish Island groups:*

Stone 3: Scottish Island groups.
Redrawn illustration based on an original Stonehenge Laser Scan illustration 11 [01]

Stone 4: *British and Irish groups:*

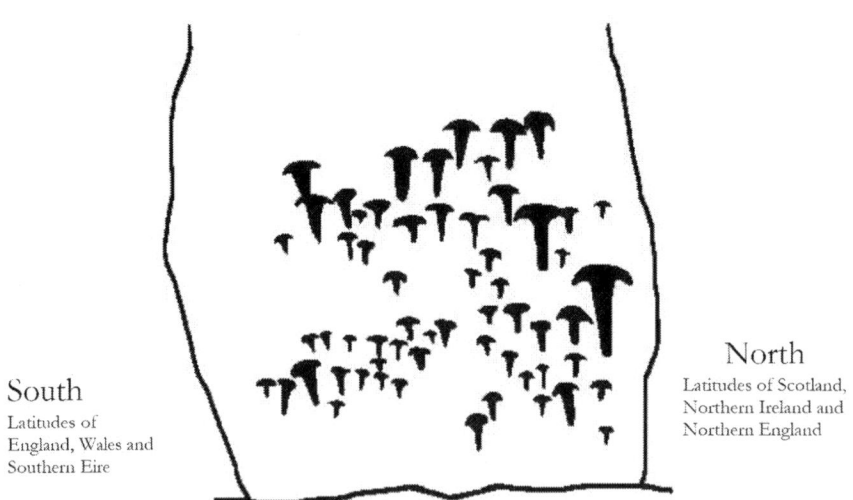

Stone 4: British and Irish groups.
Redrawn illustration based on an original Stonehenge Laser Scan illustration 12 [02]

13: THE MAKERS' MARK

Stone 5: *Continental European groups:*

*Stone 5: Continental European groups.
Redrawn illustration based on an original Stonehenge Laser Scan illustration 13* [03]

The mark on the north face of Stone 5 is particularly interesting. If these marks represented groups of people, this particular mark may have represented the Channel Islands; off northern France. These locations also correspond with other evidence about where the people came from to construct the monument (for more detail see Chapters 2 and 3). Stone 53, next to the "Three season device", also contains 'T' marks.[04] But no other stones contain such markings other than a single possible motif on stones 23:[05]

> *"These motifs occur in four key panels (the exterior E faces of Stones 3, 4, 5 and the face of Stone 53), but four motifs are found elsewhere."... "and a dagger is present on the SW face of Stone 23."*

The best guess date for the carved axes is based on their appearance being reminiscent of a flanged bronze axe with distinctively splayed edges.[06] However, no method of dating these marks exists, and no similar marks were found on any other external stones. If Stonehenge was a geocentric model, and the marks are from the time it was built, the people who built it may have come from the places identified on the stones.

Summary

Recent results from isotope analysis demonstrate that "the Late Neolithic was the first phase of pan-British connectivity, with the scale of population movement across Britain arguably not evidenced in any other phase in prehistory".[07] Evidence from the Stonehenge environs (see earlier chapters) appears to reinforce the idea that artefacts and people started to arrive from all over Britain, and Europe, to be there.

If the makers left their mark based on where they came from, the axes drawn on Stones 3, 4 and 5 are consistent with what is known about where these people came from. This interpretation is possible if Stonehenge was intended to represent a geocentric world.

Some might argue that early people could not have known about geocentric ideas. However, these ideas were well established in the ancient world. And someone must have come up with the idea first.

14: TWIN PEAKS

Neolithic monuments, described in previous chapters, were shown to correspond to the remains of what might be found if people had been searching for knowledge about the nature of our world.

In chapter 7, it was shown that it is possible to measure the size of the world using just sticks and a hill. This method works well if the hill has sea views in both directions.

In chapter 8, it was shown that a second method can be used to measure the world. This uses two or more sticks to find heights and levels. This second method needs sea views in only one direction and works best if the views look towards a sunset or sunrise. In addition to finding out the size of the world, the method can be used to confirm that it is spherical. Chapter 8 also looked at some Neolithic monuments which appear to be located to allow this second type of check to be made.

In chapter 9, it was shown that the level method could be used at the top of a small mountain, such as Cwmcerwyn in Wales, to measure the size of the world. That chapter also showed that some of the stones of Stonehenge, which were brought from the area just below Cwmcerwyn, could be used for this purpose.

It is certain that the builders of Stonehenge knew how to make a perfectly level structure (see Chapters 3 to 6). As seen in Chapter 5, the ground at Stonehenge is not level and using only the horizon would give the wrong result; yet the lintels are perfectly level. From this one fact, we can be certain that the builders must have known how to get a level surface and, perhaps as important, were using levelling equipment of some type.

Chapter 9 showed that the radius of the world is the size of a typical human foot (approx 0.263m) multiplied by 30 five times over (0.263 x 30^5 ~ 6,391km or about 3,971 miles). More precisely, the radius of our world is actually about 6,371 km (~3,958 miles). The only numbers that would be needed to remember our world's radius, using only a fairly average human foot length, are the numbers thirty and five.

The numbers 30 and 5 are a design feature embedded into Stonehenge. Chapter 9 also showed that the number 2 is also embedded into the structure of Stonehenge: its outer circle, of sarsen stones, is a scale model of our world if base 30 is used:

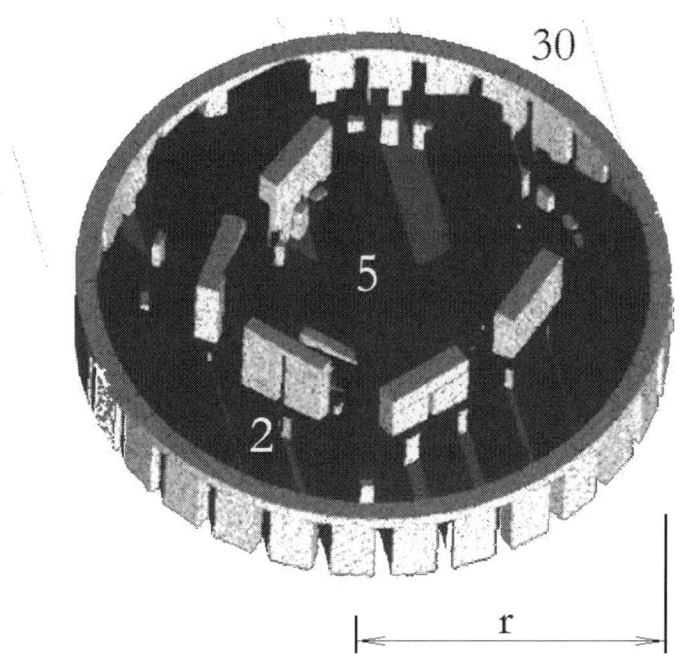

Stonehenge numbers

If the size of the world had been found long ago using levelling sticks, or perhaps trough stones shaped to suit, it may have been tempting to ask if there were better, more accurate, ways to find out the size of the world.

One very exact way of getting a precise measurement of our world's size is to use two mountains; each of a known height:

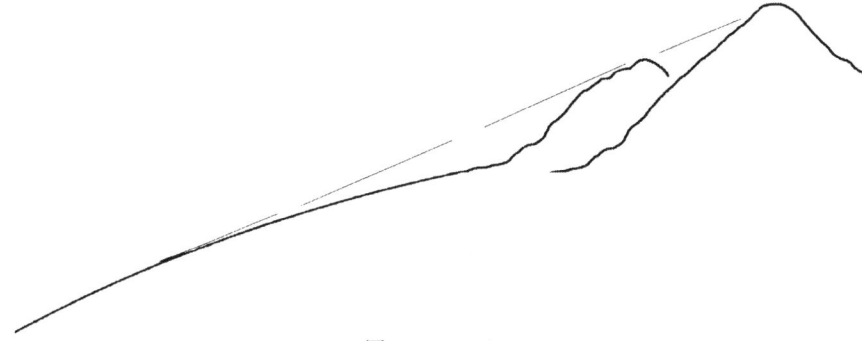

Two mountains

If the height of each mountain is measured using levelling sticks, and if the higher of the two looks over the other towards the sea, then the point at which the low mountain intersects with the sea horizon can be found by walking up the side of the higher mountain. This method is similar to using coins on the end of a levelling stick, but on a truly vast scale.

If by chance the two mountains are near to each other and have a straight and easy path between them, then the distance between mountains, and the difference in height, can be found using levelling sticks on a windless day. Providing that the eyesight of the people doing the task is very good, this method of finding the angle of slope down to the horizon would be extremely accurate.[01] The principle of this method is almost identical to the method developed by Al-Biruni in the 10th century (see chapter 7).

But these calculations become complex if the two mountains are not close. There are very few pairs of mountains which are close to each other and have a view from one mountain, over the top of its partner, to ocean.

A 'best possible case' for doing this experiment would be a hypothetical, almost magical, pair of mountains whose slope down to sea happened to be at 1:60 (1:N). From earlier chapters, the equation to work out the size of the world, using a typical human foot as the measurement unit, is:

2 x N x N x mountain height

2 x (60) x (60) x (30 x 30 x 30) / (2 x 2 x 2)

which is:

30^5

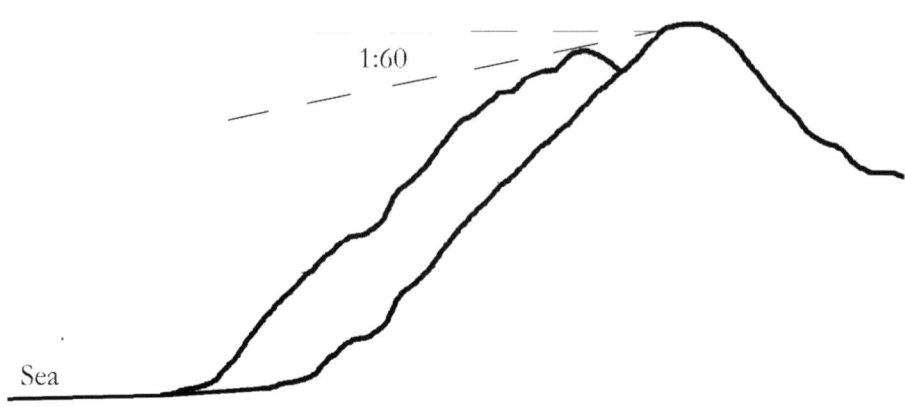

1:60 mountains

Although base 30 (see earlier chapters) is useful for finding out the size of the world, there would be no particular necessity to use it unless the tallest pair of mountains, having all of the above criteria, also happened to have a slope down to the sea of 1:60. If those conditions existed, and no other existing counting system was already in widespread use, anything other than base 30 would seem neither sensible nor suitable.

A taller mountain, enough to see a slope down of 1:30, does not exist in Britain;[02] Ben Nevis (near Fort William in the western Highlands of Scotland), stands at only 1,345 metres.

The ideal 1:60 hypothetical mountain pair would need to be approximately:

0.263 metres x (30 x 30 x 30) / (2 x 2 x 2)

This is just under 900 metres

The higher mountain would need a clear view over the other towards the sea, the mountain pair would need to be very close to each other, and the view would need to look towards a sunset or sunrise.[03]

These conditions are almost impossible to find. But if the measurement were done at such a hypothetical, almost magical, place, the mathematics needed to describe the Earth would easily resolve into just three numbers: 30, 5 and 2.

14: TWIN PEAKS

Two mountain peaks with *all* the above criteria do exist in the Brecon Beacons (of Wales). They are also the highest of that mountain range:

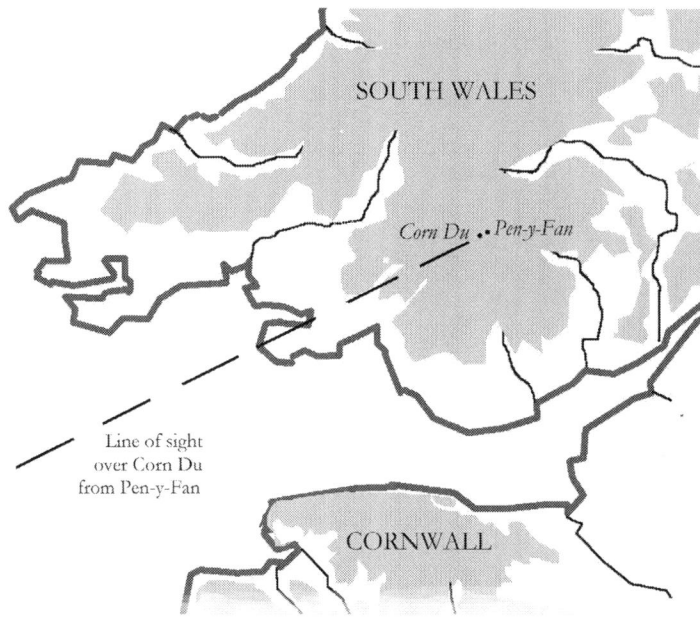

Line of sight over Corn Du from Pen-y-Fan

These peaks are known as Pen-y-Fan (+886m aOD) and Corn Du (+873m aOD). From looking at UK survey maps, there appears to be no other combination of mountain peaks that would work.[04]

The sight-line between these two peaks happens to look over lowlands to the Atlantic sea horizon. That viewing arc looks directly at those sunsets which occur over the days of the winter cross-quarters. In the image below, the azimuths of that ~7° arc are shown; with the shaded areas indicating the obscuring mountains, or hills, found to either side:[05]

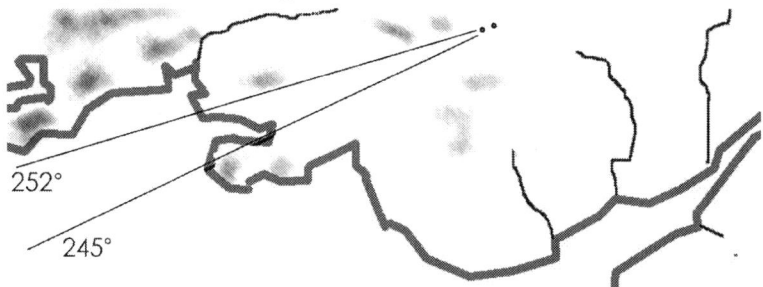

Unobscured view azimuths from Pen-y-Fan

The sight-line, from Pen-y-Fan through Corn Du, passes over the Atlantic ocean. In the photograph below, hills to the left and right of Corn Du obscure the view to the ocean horizon. In normal daylight hours, it can be difficult to see exactly where that ocean horizon is:

Corn Du, and the gap between mountains to the Atlantic, seen from Pen-y-Fan

As seen in the shot above, the pathway up to Corn Du is steep but relatively straight. In the other direction, and as seen from the path, the ridge between the two peaks forms an almost straight pathway:

The path from Corn Du to Pen-y-Fan

The distance between these two peaks is slightly more than 500 metres. However, the position at which the Atlantic can be seen to meet Corn Du occurs just below the top of Pen-y-Fan; at a distance of about 470 metres from the cairn on the summit of Corn Du:

The cairn on the summit of Corn Du

The ideal place to view the meeting of the Atlantic horizon with the cairn at the summit of Corn Du can be found at the location shown below. Here, the difference in level is about 30 human feet (or just under 8 metres).[06] This drop in level is due to the combination of the 1:60 slope (down to the ocean) and the distance between a viewing position and the top of Corn Du (about 1800 human feet or ~470m).

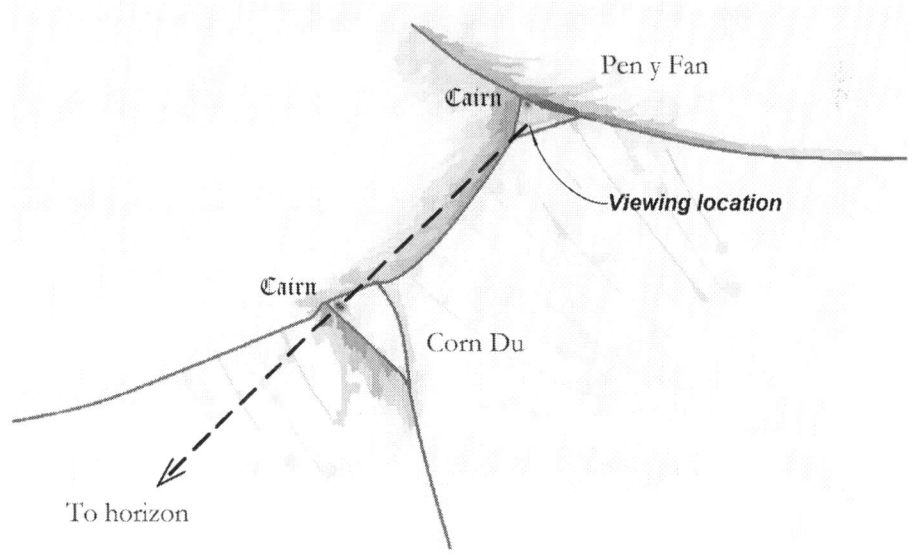

The viewing line from the Twin Peaks

The viewing locations occur just below the summit of Pen-y-Fan (see the arrow below):

Pen-y-Fan seen from Corn Du

When sunset is viewed from Pen y Fan at a winter cross quarter, the horizon will become easily visible if it is back-lit by the Sun:

Zooming in on the back-lit horizon at a winter cross-quarter

These extraordinary topographical coincidences mean that the task of accurately finding the slope down to ocean would be extraordinarily simple for anyone using a simple water level and a base 30 counting system (30 'feet'/1800 'feet' = 1/60).

At about 880 metres, the height of the mountains is also relatively simple to find, albeit a much longer task given that the nearest coast is just over 40 kilometres away. The height of this location would be measured at about 30 x 30 x 30 / 8 human 'feet'. These two sets of coincidences at this location set the numbers 30, 5 and 2 as very obvious choices with which to describe the radius of our world (30^5 human feet).

The location on Pen-y-Fan, at which the top of Corn Du can be seen to meet the Atlantic, can be found by looking for folds in the ground. At this height above sea level, it is possible to walk between those slight folds, located off the main path, and to see Corn Du at azimuths varying from about 245 to 252 degrees. It is a remarkable topographical coincidence that these azimuths correspond with both the cross-quarter sunsets[07] and the visible Atlantic alignment on the far horizon.

But no archaeological records appear to show that these folds are anything other than natural:

Ground folds just below summit of Pen-y-Fan

However, these folds in the ground do appear to be ideal markers at which the meeting, of the Atlantic and Corn Du, can be precisely seen when the sun sets at the horizon. This sunset view only occurs around the days either side of the winter cross-quarters (Samhain and Imbolc).

For anyone living at the very beginning of scientific inquiry, this place, its perfect horizon, and the way that the numbers describing our world resolve to just 30 and 5, could have seemed exceptionally special. The mathematics come together at this location to show that the radius of the Earth is 30x30x30x30x30 'feet': something that might need to be written in stone by constructing, to scale, a circular model to describe it.

Pen-y-Fan and Corn Du have relatively flat tops due to a core of Upper Old Red Sandstone overlain by a Grey Grits Formation. Lower down, and at the north side of these mountains, a bedrock of Senni formation sandstone can be found at the lower levels. This Senni formation starts about three to four hundred metres below the summits of the two peaks and is shown shaded below.[08]

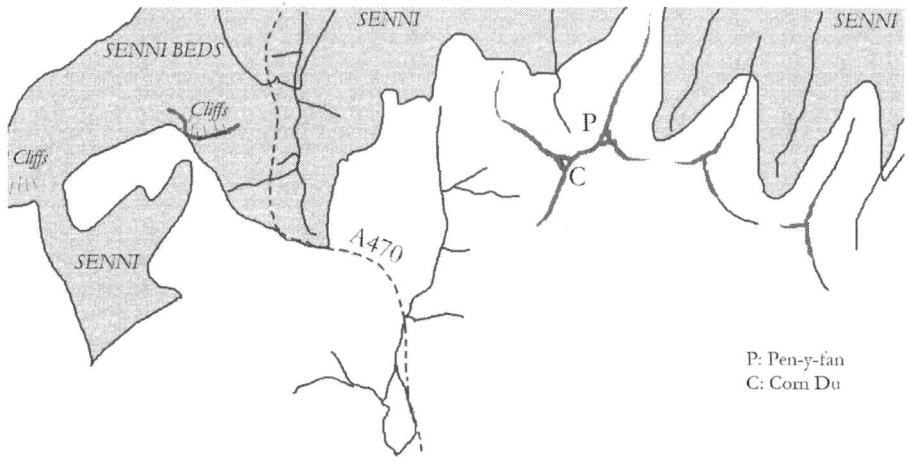

Geology close to Pen-y-Fan

Directly west of Pen-y-Fan, a massive gouge in the landscape was once carved by a glacier (location shown in image above). These are the cliffs of Craig Cerrig Gleisiad:

Looking east from the side of Pen-y-Fan

The cliffs

From those cliffs of Craig Cerrig Gleisiad, the view of Pen-y-Fan and Corn Du stands out above the horizon:

The two peaks seen from the cliffs of Craig Cerrig Gleisiad

This area, known to have been inhabited in ancient times, forms an environment sheltered from the prevailing south-westerly winds. It also contains gentle streams:

The cliffs

The cliffs are dangerous to climb due to very steep slopes and loose rock. Below those cliffs, collections of very large rocks can be found, strewn across the landscape. These large rocks are due to previous collapses of sections of the cliffs above:

Beneath the cliffs of Craig Cerrig Gleisiad

The rock which makes up these cliffs, and the massive fallen boulders below, is a Senni stone which, when cut, is a light green-olive colour. This type of light green-olive Senni stone is thought to have been used to make the Altar Stone.[09] Though the exact source of the Altar Stone's rock remains unknown (because Senni stone occurs throughout southern Wales and as far as the English borders), this type of stone may have been considered special by the people who built Stonehenge.

15: THE UNIVERSE

A geocentric universe describes a fixed world around which everything else rotates. In Ancient Greece, 6th century BC, Anaximander of Miletus first proposed that the Earth was a cylinder located at the centre of the Universe,[01] with everything else in the sky being holes set in invisible wheels that surround the Earth. The next big step was put forward by Ptolemy who, in the first century AD, proposed a mathematical scheme, with a spherical Earth at its centre, which could successfully predict the movement of planets.[02]

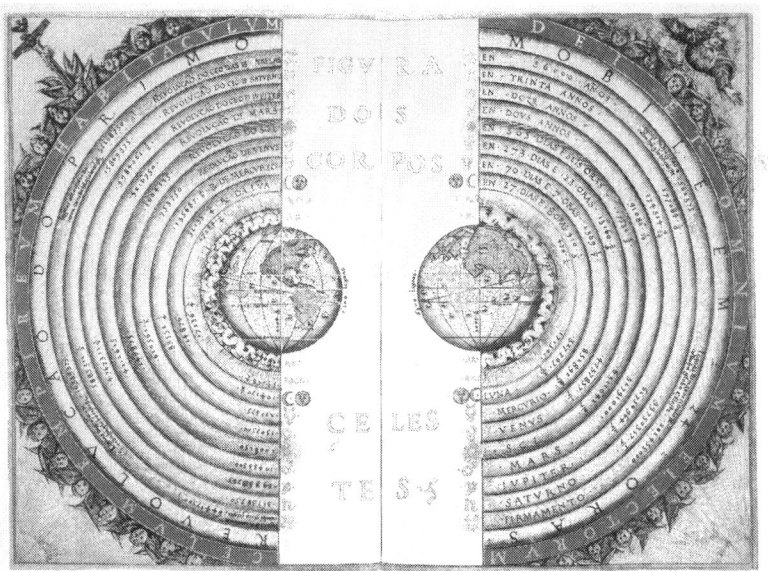

The Ptolemaic geocentric model of the Universe in 1568 Bartolomeu Velho (Bibliothèque Nationale de France, Paris). [03]

However, there were different views: Aristarchus of Samos, who lived nearly three hundred years before Ptolemy, had proposed that the Earth revolved around the Sun.[04] But long before Aristarchus, a follower of Pythagoras (named Philolaus) believed that the Earth revolved around a central fire.[05] This heliocentric theory was given short shrift by Aristotle, the leading philosopher of his day.[06]

Fifteen hundred years after Ptolemy, Copernicus re-introduced the idea that the Earth revolved around the Sun.

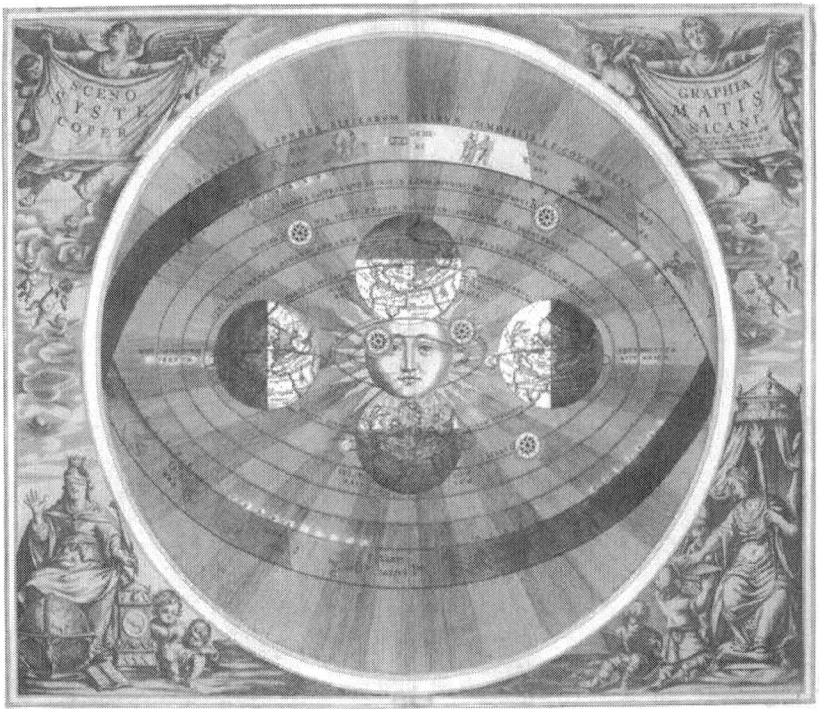

The Copernican system by Andreas Cellarius from the Harmonia Macrocosmica (1660) [07]

Copernicus' work was first printed in 1543, but the geocentric view persisted for a very long time; largely because his 'De revolutionibus orbium coelestium' (In English: On the Revolutions of the Heavenly Spheres) was considered incompatible with the Catholic faith. It was listed by decree, in 1616, on the 'Index of Forbidden Books'.

Coincidences: Part 1

In 1559, William Cuningham, in his book 'The Cosmographical Glasse', drew out the established view; showing how the Sun seems to rotate about a fixed world. In the image below, his original drawing has been faded (on the right-hand side) to allow the solstice positions to be seen:

A geocentric view: Extract from 'The Cosmographical Glasse' of 1559 [08]

Cuningham has chosen to view the world looking towards the east. This eastwards view allowed him to show both the angle of the polar axis and also how the zodiacal constellations (from the tropic of Cancer to Capricorn) intersect with the solar planes at the solstices.

Similarly, Stonehenge is built in a perfect circle and has a circular bank outside. The axis to the Heelstone, along the Avenue, is in the correct direction to show the polar axis (just *over* 51° clockwise from north). Stonehenge has four markers known as the Station Stones; which are also in the correct position to show the position of the Sun at its extremes: the two solstices.

Cuningham has shown thirty-six markers; one for every ten degrees. But other marker systems are possible: Stonehenge is one-seventh of a world's diameter from the equator. Fifty-six divides into seven precisely, leaving seven sub-division points between each marker. Within Stonehenge's bank, there are fifty six markers known as Aubrey holes.

Stonehenge's circular bank is unusual[09] in that it has a ditch *outside* the bank. By comparison, most other 'henges', such as Avebury, have an arrangement that has a ditch inside the bank. This inwardly facing bank at Stonehenge could represent the outer edge of the Universe.

Stonehenge is at a latitude of 51°10'44"; almost precisely one-seventh of the way around the world from the equator. If a line representing the equator were drawn onto Stonehenge's plan, its Station Stones are at an angle of approximately 24° relative to that 'equator'. In 2500 BC, the Sun's apparent movement was 24° either side of the equator:

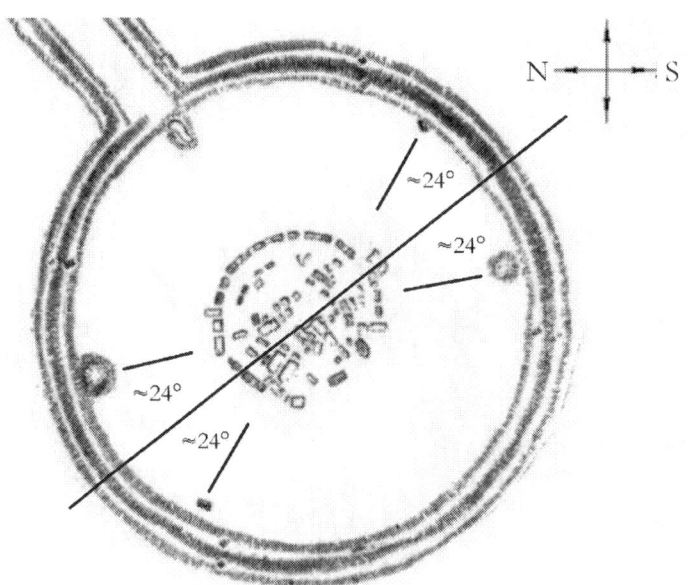

Modified & updated version of Stonehenge 1845 Ground plan

Using fifty-six degrees of measurement rather than the modern equivalent (360°), each solstice stone marker must be 3.7 of the fifty-six 'degree posts' away from the equator line (i.e. 24°×56/360°). At Stonehenge, stone 91 is approximately 3.7 Aubrey holes away from hole 14. Stone 93 is approximately 3.4 holes from hole 42. Stones 92 and 94, though both now missing, were about 3.7 holes away from holes 14 and 42 respectively.

Coincidences: Part 2

The earlier coincidences within this chapter establish that there may be some sort of existing ('a priori') evidence for a geocentric connection. Stonehenge also has an internal structure, as described earlier, which precisely duplicates a system designed to show and explain the geocentric nature of the Universe.

To re-cap, the similarities related to the internal structure are listed below. Explanations and diagrams of detailed mathematics and engineering considerations have been shown in the notes rather than in the main text:

1: Height of winter mirror rim
The base of the winter polar axis rod must be set at ground level,[11] it must be accessible and people must be able to see the 'shining star' from outside.[12-14] Either by trial and error, or by design, there is only one really good way to make this work.[15&16] And for a 32m diameter bowl, this arrangement sets the height of the outside edge at a maximum of 5.0m tall (16'5").[17]
- Stonehenge's outer sarsen ring has a height of about 4.9m to the top of its outer rim (Stones 1–30). It also has an external diameter of about 32m. However, there is nothing particularly special about this coincidence, because we started with the assumption that this is what we might be looking at: so this coincidence is just a starting point.

2: Diameter of the circular rim for the winter arrangement
Because a 32m (105') diameter mirror sphere sits above the circular rim, the internal diameter should be slightly smaller. The smallest diameter, for this size, and which fits at this latitude, is just under 30m (≈99').[18]
- Stonehenge's internal diameter is just under 30m.

3: Pure circular rim, internally facing for the winter arrangement
The lintels need sufficient depth to allow placement of the winter mirrors. They would also need to be circular on their internal faces so that the mirror frames can be placed in the correct initial position.
- Stonehenge has a pure circular rim, internally facing.

4: Level surface at top of outer lintels

To make this work easily, the mirror frames need to be positioned so that their tops are level. For the winter position, if frames are made to lean against a circular rim, they can be set to the correct height using a fixed distance (up or down) from some sort of level surface. If that is done, the mirrors have now become part of a sphere, but still need final focusing.

- Stonehenge's outer lintel stones form a precise circle with a level top surface.

5: Flat surfaces of supports

The external pillars, arranged in a circle, would require a flat internal face to allow positioning of the bottom of the mirror frames. Using a fixed length rod, these can then be quickly set to position and the bottom ends raised if required. Once done, the mirrors are almost perfectly focused.

- At Stonehenge, the external sarsens (Stones 1–30) are flat-faced on the inward-facing surface.

6: The dimensions of the sphere and the polar axis socket

At a 16m (52' 6") mirror sphere radius, the bottom of the winter pole will be located at ground level; provided the centre of the hinge (and the mirror sphere) is just under 10m (≈33") above ground level. This means that the back face of the socket must be both directly south of the centre and located about 8m (≈26') away from the centre (see note 13).

- At Stonehenge, a socket occurs at the easterly edge of Stone 54. The easterly edge of Stone 54 is directly south of the centre of Stonehenge. The back of the socket is within six inches (15cm) of the circle's centre line, going in a north-south direction, and is located approximately 8m (≈26') away from the centre of the monument.

7: Detailed dimensions of the lowest socket

The lowest socket should be oriented towards the centre (looking north) and should also point upwards. This allows a tree-pole to be brought in and then raised up to an angle of 51°.

- At Stonehenge, the lowermost socket on Stone 54 is oriented to the north and, although badly damaged, is arranged as a curious egg-shaped socket whose top surface tapers away. This would allow a tree-pole to be brought in and then raised to an angle of about 51°.

8: Platform Stones set to correct plane and location for access

Although spherical solar concentration works by focusing to a line to get concentrated light (see Appendix B), the best length for the rotating rod will be between 8m-10m (≈26'-33') for a 16m (≈52') radius mirror sphere. To allow access to the end of this focal point (on the rod), a good solution is to use thin platforms which cast the least shadow.

- At Stonehenge, a thin horse-shoe shape has been installed with four sets of short trilithons (stones 51&52, 53&54, 57&58, and 59&60). Each stone set, including the Great Trilithon (55&56), is set at a height which allows access to the end of the rod for each of the three levels (see 11 below for explanation and notes on the 'three-season' levels).

9: Trilithons set to correct locations for hauling

To make an effective hinge, some sort of platform arrangement is also needed against which the sail can be rotated (around the tree-pole) using ropes. These platforms must not cast too much shadow on the mirrors at any time of year. For this reason, any hauling platforms near the mirrors should be the shortest.

- Stonehenge has five sets of internal trilithons (Stones 51&52, 53&54, 55&56, 57&58 and 59&60). In addition to appearing to be configured for access, these sets also double-up as hauling platforms and appear to be ideally configured for it (see tests done in Appendix B). They are also arranged so that they cast the least shadow: the shortest trilithon pairs (51&52 and 59&60) are at the front.

10: The Great Trilithon

Stones 55 and 56 (the Great Trilithon) are 7.5m tall. However, if Stonehenge were a geocentric concentrator, the current position of the remaining Great Trilithon stone would be better sited half a metre or so towards the north-east.

- Recently, it has been discovered that this trilithon stone was re-positioned by the Victorians: it was moved from the north-east.

11: The upper sockets

As the Sun's plane goes up with the approach of summer, the plane of the focal circle (of the equipment) goes down. To adjust for this, the rotating rod must be tilted downwards. To get the focal point to stay at approximately the same height above the lintels throughout the year, the equipment must be raised so that the rotator can be hauled from the same set of platforms. This happens once for equinox and again for summer. The mirrors can be tilted and re-positioned (see also 13 below).

The solar plane varies in a sinusoidal motion, which means that the position of the solar plane is in the bottom 15% of the range in winter, then from 15% to 85% during spring, and the top 15% in summer. If the equipment were raised, the ideal [19] vertical distance between the top and the bottom polar socket will be about 4.2m (13'9').

- At Stonehenge, there appear to be two further sockets on the edge of Stone 54. These are in a vertical line within that one stone. The dimension from the lowest to the uppermost socket is about 4.2m (13'9'). Although the sockets appear damaged, perhaps from centuries of use, each of these sockets appears to be correctly orientated to receive a tree-pole placed at an angle of 51°.

Measuring at Stonehenge

12: Special (unusual) packing for the 'Socket stone'

This stone, being the only stone which is required to have tree-poles raised against it, needs to have special foundations.

- Stonehenge's Stone 54 has unique packing, is the heaviest stone of all at Stonehenge, and has an unusual bulbous footing. This stone is also the correct size to accept the huge lateral forces generated by raising a tree-pole. [20]

13: Spring & autumn mirror supports: Height

The secondary mirror supports have to be arranged in a circle, particularly about the north-east. These support stones must also be the correct height and location. This positioning all depends on the location of the centre of the hinge when the tree-pole is placed in the second socket.

For the spherical demonstrator, with an internal diameter of some 30m ($\approx$100'), a secondary support ring at the north-east, approximately 23m to 24m (75' - 79') in its plan diameter and some 2m (6'6") high, would suit the task (see illustration in notes).[21] Note that other diameters are possible providing the height of support is modified to suit.

- At Stonehenge, the north-east blue stones are arranged in a circular arc approximately 23.5m (77') in diameter *and* about 2m (6'6") in height.

Support stones for the equinox positions carry mirror frames and would therefore need to be strong. Location and setting can be done using fixed lengths measured from the larger finished stones of the outer rim, so these stones would need only a rough finish.

- Cleal et al. describe in detail how the Bluestone Circle stones, that is the circle of upright 'bluestones' just within the outer 'walls' of Stonehenge, are heavy and roughly made compared with the innermost bluestone horse-shoe.

14: Smooth faces on shorter Platform Stones

The shorter platform stones (including the 'Socket stone') would, ideally, be smooth internally (especially to prevent jarring as the sail is raised). From the tests done in 2013 (see Appendix B), we discovered that smooth faces also make an ideal back reflector which can be used to precisely focus mirrors once they are placed in their initial position.

- At Stonehenge, and with the exception of the Great Trilithon, all of the Trilithons have smooth internal faces.

15: Early configurations

The Q&R holes at Stonehenge appear to be in the correct position for a slightly smaller structure designed to do exactly the same thing as the later structure. There is also evidence that some of the bluestones were originally lintels (see Chapter 3: 'The Stones: A potted summary'.)

16: Internal timer ring (inner horse-shoe) in correct location
If a plumb bob at the back end of the sail is dropped to a set of markers, the position of the sail can be judged with time. This might be important on a day that might prove cloudy. If this existed, an elliptical or semi-elliptical ring of stones would be an ideal arrangement; and this ring of markers could only be constructed within the larger trilithon horse-shoe.

- The bluestones of Stonehenge's inner horse-shoe are (by comparison with the bluestones near the external rim) well tooled, polished, elegantly composed, shaped, and neatly arranged. They are also in good positions to act as a timing ring (see illustration in notes). [22]

17: Instructions on how to put up the pole and sail
To make it absolutely clear to anyone setting up the pole, a drawing of the sail and pole might be carved into the side of the 'Socket stone'.

- At Stonehenge, at the edge of Stone 54 (the stone with the sockets), there is a T-shaped relief carving. It is known as the 'Chief's Face' because it looks a little like a representation of eyebrows and a very long nose. The 'T' also has the correct relative dimensions to show the pole, its sail and the counterbalance distance necessary. [23]

Stone 54 (photo courtesy of Terence Meaden)

18: T shapes
At Stonehenge, other large T shapes appear to exist on the back of 53, 54 and the side of 53. From recent scans, large numbers of additional T shapes also exist on the front face of stone 53 (see Chapter 6: The 'T' Rune.)

19: Location
On a geocentric world, drawn to show the axis of the North Pole at Stonehenge's location, the position of Stonehenge itself is at the top of that image of the world. This position is directly east of the centre (see also illustration in notes). [24]

- From recent laser scans, almost all remaining T shapes cut into the stones are on those stones directly east of centre.

20: Viewing
The arrangement would ideally be viewed from the best possible angle. For a concentrator designed to be seen in the afternoon, with the morning used to set the arrangement up (and check that it all works), this direction is from the north-east along a long strip of ground.

- At Stonehenge, recent laser scanning has revealed that the monument was specifically designed to be seen from the north-east. The Avenue, a wide strip of land coming from the north-east, would have been ideal for viewing.

21: Metals
A large amount of archaeological evidence backs up the idea that metals were in use at Stonehenge before it was built. If mirrors were in use, they would have been exceptionally valuable, so unlikely to be left at the site to be discovered later. Nevertheless, a tablet of tin was found at Stonehenge with, on one side, an inscription in a lost and unknown language (see Chapter 4; 'Antiquity and Stonehenge').

22: Legend
If such a thing were done, there would be expected to be legends passed down through generations. In the case of a solar concentrator, several of those legends appear to exactly describe what was done (see Chapter 12).

23: Remains of knowledge
If the meaning of the heavens were known at Stonehenge, there might be expected to be the remains of experiments showing how this was known. There do appear to be large numbers of such remains, all configured as required, along the South Coast of England and Wales (see Chapters 7 to 9 and 14).

24: The meaning of Heel
The word heel is derived from the meaning 'to rotate'. The word 'Yule', meaning the time of winter solstice, is also thought to derive from this word.[25]

- At Stonehenge, the stone with this name is the same as the stone about which the Universe would be seen to rotate. (for more see Chapter 3: 'Potted history: Stage 2 to 5'.)

25: The meaning of Stonehenge
The words Stone and Henge are remarkably similar to all North European words for stone, hinge, tin, and angle. These words happen to describe the main features of the solar concentrator mechanism (see Chapter 2: 'Stonehenge: The word'.)

In short, there is nothing about Stonehenge which suggests that it could not have been capable of demonstrating the nature of the Neolithic Universe. Other suggestive coincidences related to other monuments, and the source materials (stone) for Stonehenge, are detailed in other chapters, but are too numerous to summarise here.

Summary

The features described above may just be coincidences. Nevertheless, they appear to describe every aspect of the design at Stonehenge, and other monuments, on these islands.

The British Isles have long been fascinated with magic. Authors such as Tolkien and Rowling have brought new life to old folklore.

Tales of magical wands, shields, bright lights and mysterious powers are all recorded in British and Irish legend. Getting tin and other metals from stone, a process which requires a stream or lake, would have been the first step in the magical process of making a mirror.

When Leonardo da Vinci produced the Last Supper, one of the most reproduced religious painting of all time, it is said that people would watch the drawing for hours:

The Last Supper (public domain version)

This was the reaction of people familiar with the new sciences of the 16th century. Although familiar to us now, the painting used the new idea of perspective to focus on the central figure and also showed 'real person' reactions of the disciples.

But the people of the Neolithic may have only just started to imagine science. A bright daytime 'mini-sun' created on Earth would have had an immense impact on how they saw their place in the world.

Stonehenge's plan layout is the same as an idealised geocentric description of the Universe. Its inner stone monument is capable of producing a spectacular public display of solar movement. The arrangement of this system appears to be based on a simple method of tracking celestial objects.

Perhaps Stonehenge was a depository of knowledge about the Universe. A place of learning, it could also have been the primary source of Arthurian stories of magic and its power. Some legends of magic perhaps remain only as fragments of the Old World's knowledge?

Greek legend tells of a Titan named Hyperion: [26]

> *"Of Hyperion we are told that he was the first to understand, by diligent attention and observation, the movement of both the sun and the moon and the other stars, and the seasons as well, in that they are caused by these bodies, and to make these facts known to others; and that for this reason he was called the father of these bodies, since he had begotten, so to speak, the speculation about them and their nature."*

And of a mysterious people known as the Hyperboreans who lived in the north on a very large island somewhere beyond the Mediterranean: [27]

> *"There lies in the Ocean an island no smaller than Sicily. This island, the account continues, is situated in the north and is inhabited by the Hyperboreans, who are called by that name because their home is beyond the point whence the north wind (Boreas) blows ... And there is also on the island both a magnificent sacred precinct of Apollo and a notable temple which is adorned with many votive offerings and is spherical in shape ... The account is also given that the god visits the island every nineteen years, the period in which the return of the stars to the same place in the heavens is accomplished."*

If England was Hyperborea, and Hyperion also represents those people, then Stonehenge's lost purpose; to store and save ancient North European knowledge about the nature of our Universe; was only rediscovered thousands of years later.

16: EPILOGUE

If the stone monument was a geocentric description of the Universe, it specifically focuses on the movement of the Sun, at its solstices, more than any other astronomical feature.

Evidence suggests that the original monument, before the stones were placed, was originally aligned to the solstice sunset or sunrise. Some say that this was the sunset at mid-winter (approx December 21) whereas others say it was the dawn at mid-summer (approx June 21).

At the time in which the first phase of the original monument was constructed, huge swathes of land were being set aside for the construction of monuments such as barrows and cursuses. At some point before Stonehenge was built, these constructions were abandoned.

Some believe that Stonehenge, the new monument of stone, enhanced the solstice alignment in some way. But, if you agree with this book, you may also agree that it probably does not.

Whatever the reason for the early solstice connection, the motivation to construct the stones is likely to be connected to whatever had happened in the times before. If Stonehenge was all about knowledge, explanations for what happened before should be simple, logical, and obvious.

Many people believe that Stonehenge was a temple devoted to the worship of the Sun. Though the Sun was obviously very important, the ideas in this book suggest that something else had been going on. Something older, and more important to them, than honouring a God named Apollo.

The Universe described by a 'Stonehenge' would help to heal the fears of the old times. Some of those ideas would be set in stone to become a permanent reminder of the new beliefs. But as time went by, the new ideas may have become an unquestionable doctrine. If that happened, the chambers of many of the structures, designed to question the nature of our world, would have been blocked up. This seems to have happened at places such as West Kennet:

West Kennet: The blocking stone placed at the entry to the passage

Thousands of years ago, people feasted at the time of Yule. They may have exchanged presents. Over the winter solstice, they may have celebrated their beliefs by congregating around a tree-pole filled with light.

In the current age, people feast at the time of Yule. They exchange presents. During the time of the winter solstice, some celebrate by congregating around a tree filled with light.

But the fears of climate change, which this book suggests are described in some ancient places, have re-emerged in a new but familiar form.

APPENDIX A: The rotation of our planet

The Earth rotates a full turn every day, which gives the appearance that the Moon and the Sun go around us.

Our planet also moves in a big circle around the Sun, but our North Pole always points to the North Star. This star is at right angles to our direction (or plane) of spin, but is not at right angles to our orbit around the Sun.

3,000 years ago, the 'Guardians of the (North) Pole' were the stars Kochab and Pherkab. Polaris, known to the Greeks as Phoenice, was just an ordinary star at the other end of the Big Dipper.

Every 26,000 years or so the celestial pole turns in a full circle with a radius of some 23.4°. Alrai (Gamma Cepheii) will become the Pole Star in a thousand years followed by Alderamin (Alpha Cephei) in 7500AD. The role passes to Deneb, Alpha Cygnis, in 9000AD and then Vega, opposite to Polaris, in 14000AD.[01] Our planet, like most others, 'tumbles' in space so slowly that we have the same pole star from year to year.

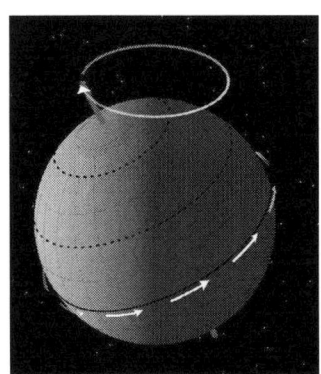

Precession: Image courtesy of NASA

Our planet rotates about the current celestial polar axis once a day. Every 365 days or so, we also orbit the Sun in a squashed circle known as the Elliptic. Our orbit also has its own polar axis, which is 23° or so away from the Earth's celestial polar axis. Another term; the Ecliptic, describes the apparent path of the Sun around our world. To avoid confusion, this book sometimes refers to this as 'the solar plane'.

Night and day are caused by the daily rotation of the World about its polar axis. Winter happens, in the Northern Hemisphere, when the North Pole faces away from the Sun (by the 'tilt' of 23.5°). Six months later, around the other side of our orbit, the North Pole faces towards the Sun giving the Northern Hemisphere its summer:

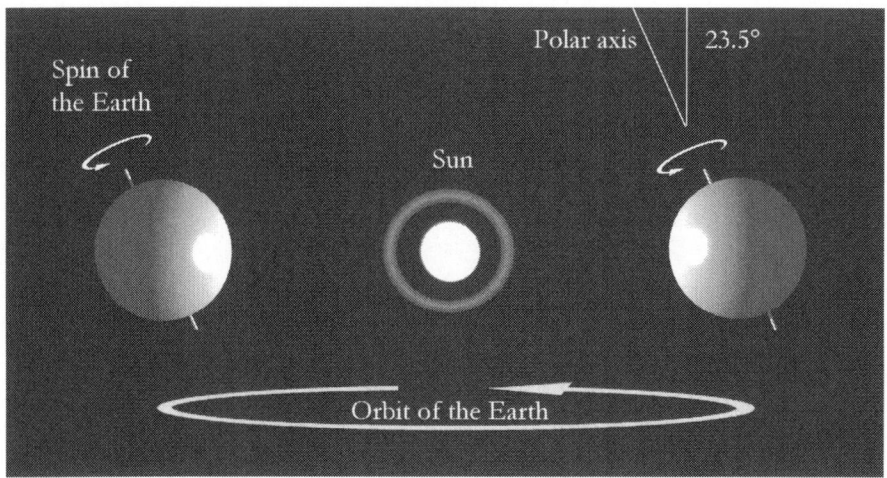

The Earth circling the Sun

Our 'tilt' is currently 23.5° away from the plane of our orbit. However, even this changes over time: over the last million years or so, it has varied between 22° 02' 33" and 24° 30' 16", with a mean period of 41,040 years.
[02] In 3000 BC, it was close to 24° rather than the 23.5° we have now.

APPENDIX A: The rotation of our planet

Some technical stuff

At higher latitudes, we need heat most in the winter and least in the summer. Because the Earth is rotated away from the Sun in winter, its rays fall at a shallower angle; so fewer watts of energy hit each square metre of soil. In addition, the Sun's rays must travel through a greater distance of atmosphere.

At the edge of the atmosphere, the Sun's energy density is 1,412 W/m² in early January and some 1,321 W/m² in early July.[03] Our orbit brings us slightly closer to the Sun in the winter resulting in the Northern Hemisphere facing away from the Sun when it is at its 'warmest'. This effect gives the Northern Hemisphere slightly less extreme temperatures than the Southern Hemisphere.

The Sun, a yellow dwarf, emits heat and light (black body radiation) at around about the same wavelengths detected by our eyes. Of the radiation that hits our planet, the highest frequencies are largely absorbed by the outer atmosphere so that, at the top of mountains, the energy density rarely exceeds 1,100 W/m².

At ground level along the Equator, when the Sun is directly overhead during the spring and autumn equinoxes, the midday solar energy density perpendicular to solar rays can reach 950-1,000 W/m².[04] At this time and location, the energy only travels through one atmosphere's 'worth' (known as an 'Air Mass'). At the start and the end of the day, the rays must travel through a greater distance; so a greater proportion is reflected or absorbed by the air.[05]

The further the light has to travel after striking the upper atmosphere, the more of the remaining ozone filtered light is reflected by air, dust, and clouds. At the Equator, a greater proportion of the high energy radiation gets through; the sky is a darker hue of blue during the day than it is at higher latitudes. Clouds also reflect light and can drastically reduce the amount of radiation getting through.

However, clouds also deflect and re-reflect light. In addition to the direct brightness which causes shadows (known as 'In Beam' irradiation), some additional light will arrive due to scatter from molecules and Aerosols (dust, water, pollutants) within the atmosphere.

Apparent solar planes

Our celestial (polar) axis is currently at about 23.5° to the axis of our orbit around the Sun, but five thousand years ago it was nearer to 24°. This resulted in the Sun being higher in the summer and lower in the winter. Warmer summers, but colder winters.

If we consider the Earth as fixed and we stand at the equator to watch the Sun move over a year, the Sun would appear to have an odd orbit which spirals from north to south. If we recorded the daily position of the spiral over a year, we would notice that it has a sinusoidal pattern which lingers over the North and South Poles during the solstices. The spiral then passes fairly quickly overhead during the equinoxes of late March and late September.

We would also record that the Sun moves 23.5° to the north in June, back to zero at the equinox and then 23.5° to the south for December.

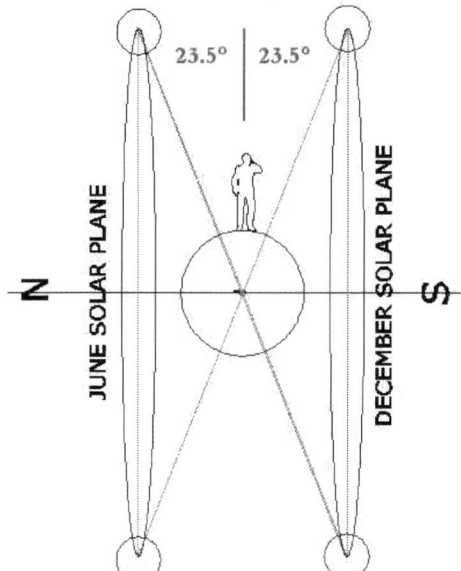

Movement of the Sun seen from the Equator

All of the above effects are a result of the celestial axis being rotated at 23.5° to the axis of rotation of our orbit around the Sun. The daily movement along this spiral is in a nearly flat circle known as the Solar Plane (or Ecliptic).

APPENDIX A: The rotation of our planet

If we move to Northern Europe and again consider the world to be fixed, our solar plane instead moves overhead during the summer; resulting in longer days. In the winter it appears to move south; resulting in shorter days.

As at the Equator, we also would observe that the annual movement is sinusoidal. This results in the Sun spending longer periods of time at very shallow angles in winter and at very high angles (overhead) in summer. We also see relatively quick changes of this angle over the autumnal and spring Equinoxes.

The man in the picture below, at a northern latitude of about 51°, sees the 'equatorial sun' (during equinox) at an angle of 39° up from ground level (90°-51°). This gives a maximum sun angle of about 62.5° in summer (39°+23.5°) and results in short shadows. In winter he sees shallow solstice angles of up to about 15.5° (39°-23.5°), resulting in the long shadows of winter.

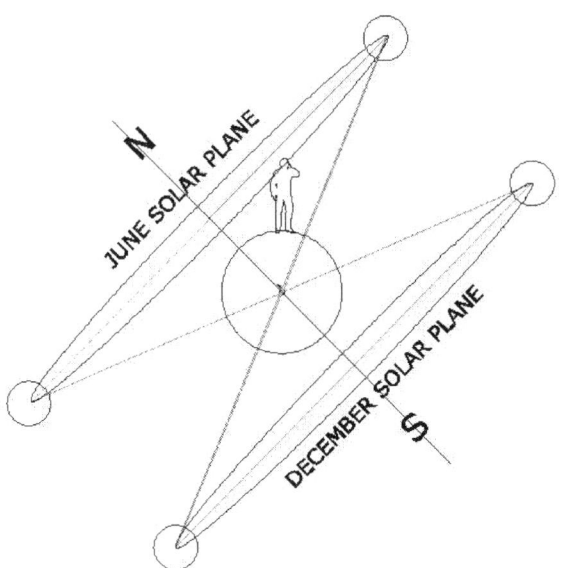

Solar Planes; movement of the Sun seen at temperate latitudes

Because we spin quickly around our polar axis every 24 hours, the apparent daily movement of the Sun is a nearly flat circle, a little like to the wheels of a bicycle, with the spokes representing time. As we move around the Sun, this circle, the 'solar plane', moves slowly up and down the polar axis.

In the picture below, the wheels represent the apparent path of the Sun around our planet, as seen from Stonehenge during the winter and summer. If the wheels were turned to face south, an ant on the ball at the centre of the picture would see that the wheels have exactly the same path as the Sun on the days of the two equinoxes:

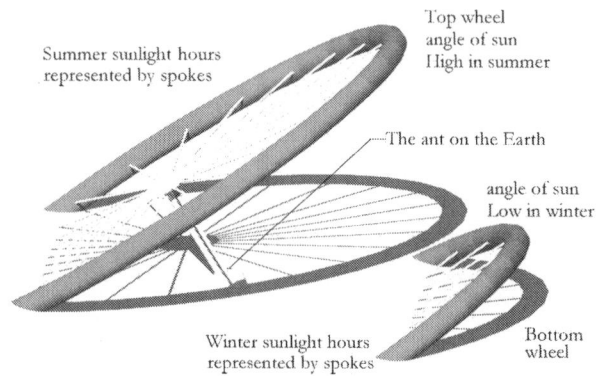

Information on how the Sun's rays affect Britain has been catalogued and plotted by the Chartered Institution of Building Services Engineers, in their excellent 'Design Guide A':

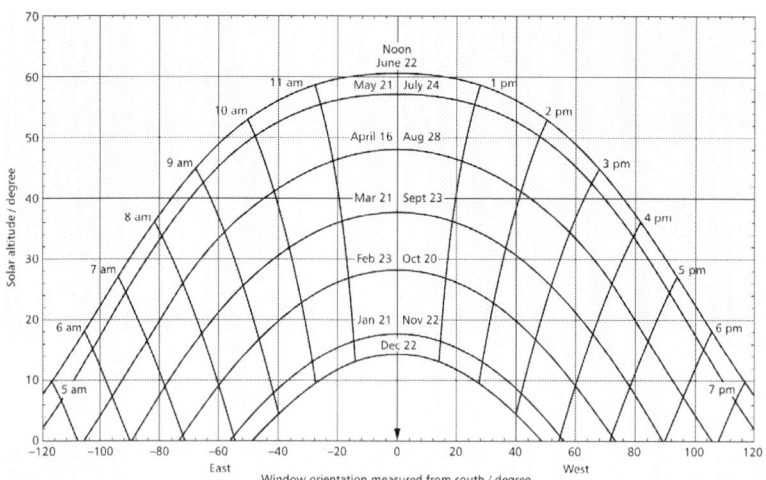

Solar Azimuths: Extract from CIBSE Design Guide A: Figure 2-12 (reproduced with permission)

APPENDIX B: Spherical solar collectors

If a set of mirrors are arranged as a sphere, rays of light are reflected from the mirrors in the way described in this book.

Spherical solar collectors, unlike parabolas, focus to a line rather than a point. However, a lot of the energy from the mirrors is focused to a point located at one half of the radius of the spherical mirror-bowl. The mirrors of a sphere only work as an effective parabola when they are within, or less than, thirty degrees of the source direction (i.e. sunbeams). If needed, all of the energy from a selected set of spherical mirrors can be made to focus to an effective point for periods of up to four hours.

On a daily basis, the mirrors need to remain fixed in position. This is particularly necessary if low-quality frames and materials are used. In summer, when the Sun is high, the focus of the mirrors is located at a low circle going round a half-sphere. In winter, the focus is higher.

If mirrors were the most expensive thing in the world, and spherical solar collection was the only known way to collect concentrated solar power, it would make sense to rearrange the mirrors to get the best efficiency. In this book, this method of arrangement is known as a 'three-season device'.

Background

Many decades go, other solutions using spherical solar collectors were proposed. Spherical mirrors, as opposed to parabolas, concentrate to a line parallel to the rays of the Sun rather than a point. However, within certain ranges, the array approximates to a parabola and can focus to a point.

Several patents using this principle were taken out during the 1970s. These solar collectors use a fixed spherical mirror bowl which concentrates the light to a line which lies between 0.5 radii and the face of the bowl.

The idea was that the arrangement would pivot and also rotate so that it aligned towards the direction of the Sun. This arrangement has the advantage that the mirrors are fixed and therefore inexpensive to build.

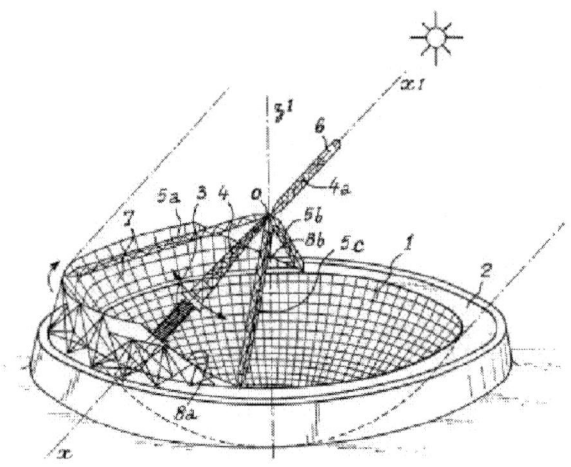

Extract from US patent no 4170985

A version of this idea had already been implemented at the Arecibo Observatory, near Arecibo, Puerto Rico. This was originally planned to be a parabolic dish until Ward Low of the Advanced Research Projects Agency noted that one other form of energy concentration could be fixed. The 'hemispherical bowl' concept, which resulted from the subsequent collaboration between Air Force Cambridge Research Laboratory and Professor William E. Gordon, allows the focus, rather than the bowl, to be moved so that it stays aligned with stellar objects.

Theory

Spherical solar collectors focus to a line. However, the distribution of intensity of ray radiation along the line is not constant: the rays tend to concentrate towards the centre of the bowl at the half radius point.

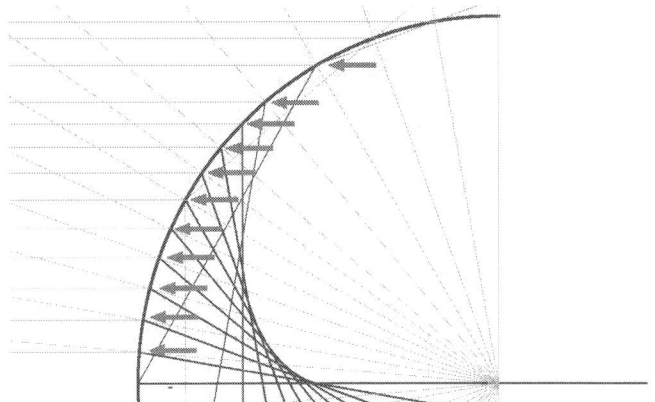

Ray reflection lines from a spherical mirror

For fixed mirror bowls with a moving solar object, only the central area of the bowl mimics a parabola. This central area continuously changes as the Sun moves.

Zooming into the half radius focal point, a line can be off-set so that a greater number of rays can be collected (shown as positions A and position B below). At a strike angle on the mirrors of up to 30° or so, the incoming reflected light can be concentrated to a very small focus area.

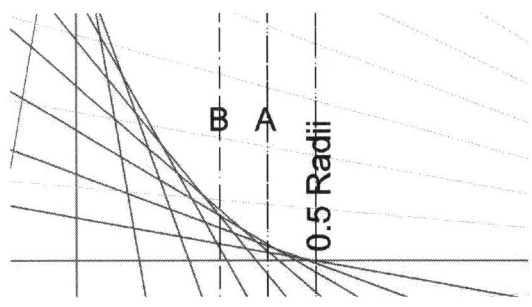

Zooming in to the half-radius point

In addition, if the reflector at the off-set is made larger than required, the sphere can be composed of flat square segments of mirror rather than a pure sphere.

The end of this focus (the half radius point) is effectively invertedly mirroring the path of the Sun around the Earth. During winter, the endpoint on the surface of the half radius focal sphere traces a high circular plane. The diameter of this circle slightly expands slightly during autumn (or spring) and then slightly contracts again during the summer.

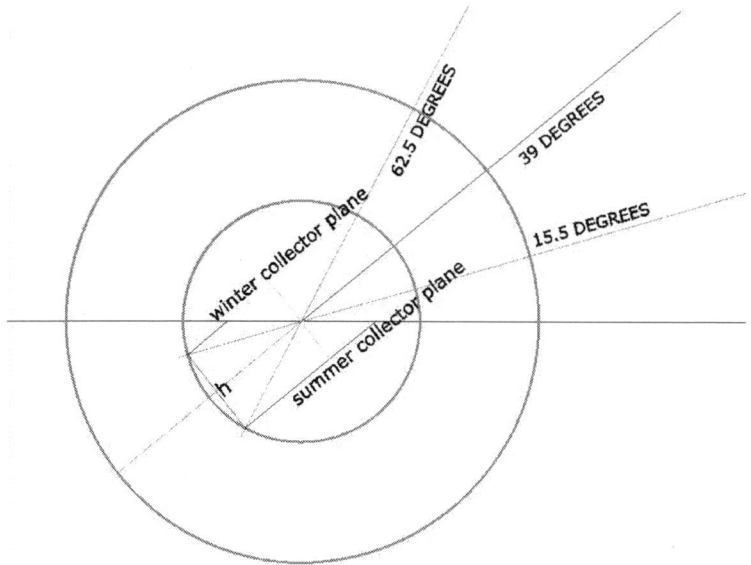

If the apparent orbit of the Sun were a pure circle, this movement (of the reflector focus point) will exactly mirror the Sun's path, showing how it moves if the Earth were thought to be at the centre of the Universe.

The above considerations show that a small flat plate, if moving in a circle and at a constant speed (similar to a 24 hour clock), will concentrate incoming light rays from a spherical mirror set.

If we decide that we are only going to collect the most productive light, the number of mirrors can be cut down to include only those mirrors that will provide concentrated light to the collector (or reflection plate). This seasonal range of daily movement translates to a 'slice' taken through the sphere of mirrors. So the most useful mirror arrangement would be a slice of a sphere focusing onto a moving 'eye'.

Consider the movement of the 'eye' as it turns in a circle. The collector will 'sweep' over the surface of the spherical mirrors as the Sun moves around its daily solar plane:

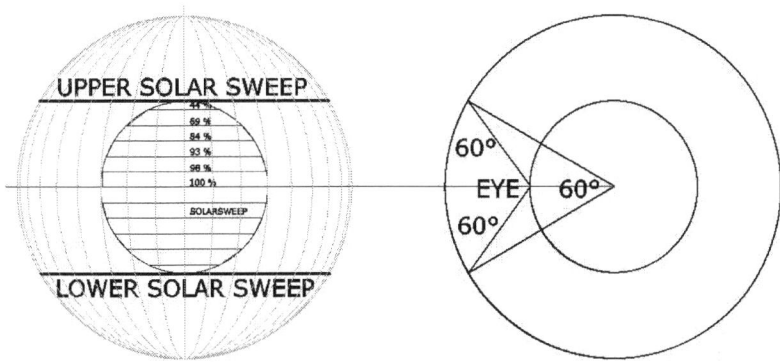

Equinoctial solar sweeps of spherical mirrors

If the axes are rotated to the winter and summer planes (23.5° either side of equinox), 'mirror bands' are obtained from which concentrated sunlight could be captured:

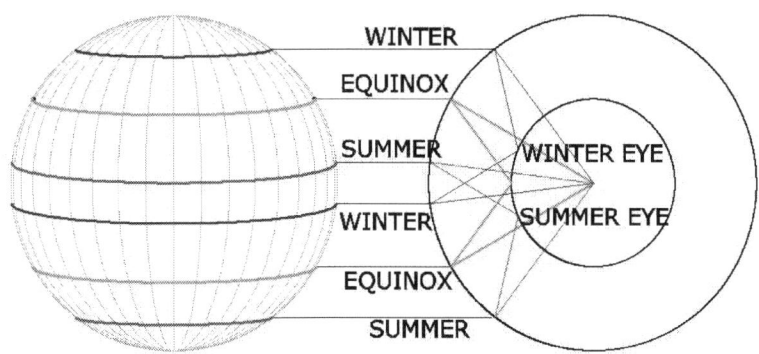

Seasonal solar sweeps

These bands now show the best position of mirrors, if they are arranged to capture this light, using spherical methods.

By rotating the arrangement to suit the latitude, the best mirror arrangement can be found for any chosen period of time. At high latitudes, we need heat most in the winter, so the ideal mirror-bowl would be at a steep angle:

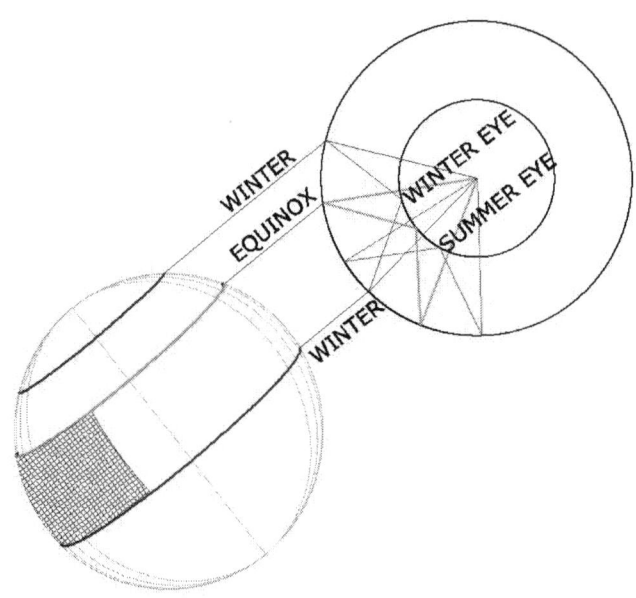

Ideal mirror zones for winter (6 month) period

In his book; 'Sustainable energy without the hot air', David MacKay, previously chief scientific adviser to the UK Department of Energy and Climate Change, looked at the possibility of paving some 5% of the UK countryside with solar panels.[06] One disadvantage of doing this is that the land cannot then be used for agriculture.

One alternative is to gather heat using widely spaced rows of *vertically* arranged spherical concentrators. This allows the land in-between to be used for agriculture during the three non-winter seasons.

It is possible that this sort of system may be necessary at some time in the future. If so, it is also possible that people of the past may have provided future generations with a method of solving their energy needs.

Before Stonehenge

During 2008 and 2009, a number of frames were built to test out the principles of spherical solar collection.

At halfway between the polar centre (the hinge) and the mirrors, light was found to concentrate as predicted. The light will remain concentrated if the receptor is rotated continuously along an arc. After testing this using a very simple arrangement (a piece of timber with a car radiator stuck on the end), I built a second test arrangement out of plywood using larger mirror sets:

One of the early second stage proof tests (2008)

The car radiator was now enclosed with polystyrene and had two flexible pipes attached. A small pump circulated water to a household storage tank, and temperature sensors allowed me to measure how much heat was being gathered per hour.

For more detailed testing over autumn 2008 and winter 2009, the third and fourth series used a fabricated roller assembly set on a Virendeel steel arc. These were placed on the same timber frame, set to our latitude (51°), and moved up and down according to the season.

Stage 3 and 4 tests

The roller sets were moved mechanically around the arcs by a small motor device; in the same way that a clock hand is moved around its face.

The roller assembly

Constructing and focusing the mirrors

APPENDIX B: Spherical solar collectors

The measured rate of collection in winter equates to about one kilowatt hour per day per square metre of mirror surface. This varies depending on the collection temperature. For instance, if collected at 80°C, the losses will be slightly higher than if collected at 50°C.

The principle of a rotating pole, moving within inwardly facing mirrors, resulted in all sorts of scheme designs to see how, and if, this could be used: equatorial solar power plants (1), near-equator heat units (2), low-cost portable cooking units (3), rigs for heating buildings in temperate zones (4), low cost drying and communal units (5) and simple portable steam sterilisation units (6):

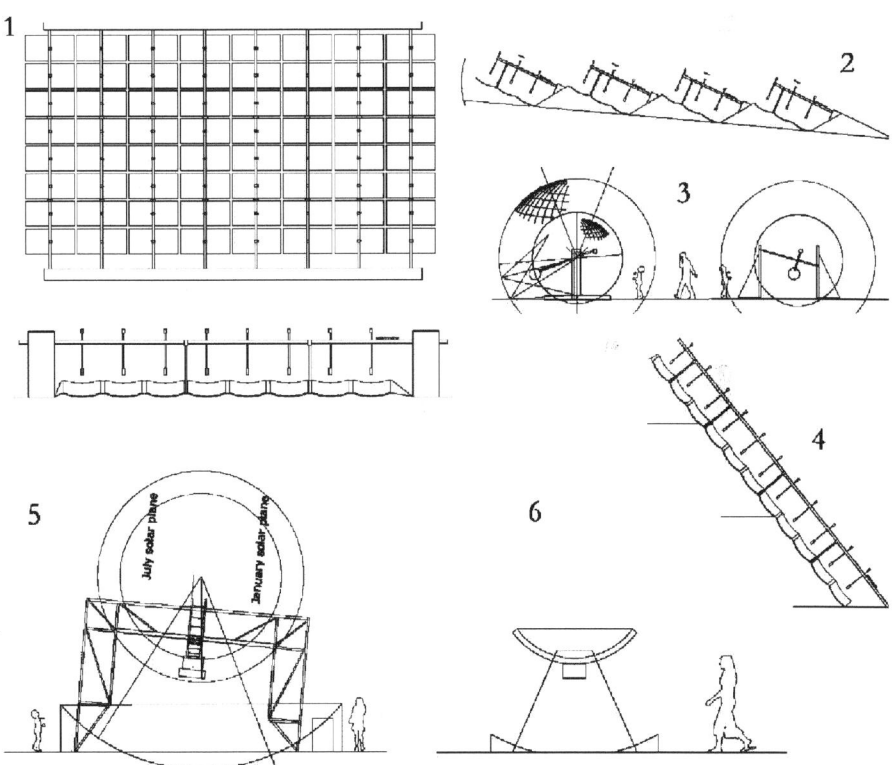

Preliminary scheme drawing extracts

Then, in 2010, whilst on a trip to visit my uncle in Salisbury, my boys wanted to visit Stonehenge. When we got there, I noticed that all the features of Stonehenge seemed to be the same as one of the forms of the later designs. So I built a computer 3D model of Stonehenge and then fitted the design in to see if there was any correlation. It was a precise fit:

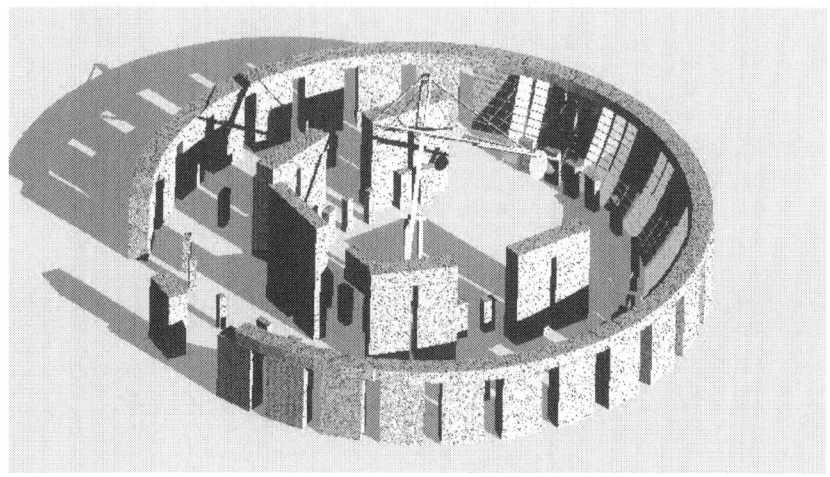

But the computer model predicted that there would have to be holes in the edge of a stone known as stone 54. So, with some trepidation, I booked a 'private viewing' of the inside of the circle.

Three holes existed, exactly in the locations predicted by the computer model:

Stone 54 at Stonehenge

I then modified the arrangement back to a pole and rod, the very first thing I had started with, and fitted a reflector:

Stage 5 tests: 2013

It was not surprising that this produced a 'miniature sun', so I duplicated the arrangement of Stonehenge in full to see if there was anything that could be learnt:

This new arrangement was a series of timber frames, set out at about 30% scale, and arranged so that the pole would be in the equinox position:

Test frame 6 from front: 2013

Each frame was tied using rope and pegged to the ground. The mirrors were then set out to the 'blue-stone' positions:

Test frame 6: 2013

The full arrangement clad in cloth:

Stage 6 tests: Willingdon, June 2013

Stay ropes were found to be necessary from 51/52, 57/58, and 59/60 to cover the full range of movement:

Stage 6 tests: Stay ropes seen from below, June 2013

Stay-rope trajectory for lowest set of the pole

The flat surfaces of the inner trilithons were found to be very useful when trying to get the mirrors focused. Lozenge type patterns form at these surfaces if the focus of the mirror set is not precisely correct:

Back-reflections: Stage 6 tests: Willingdon, June 2013

After testing, the equipment was packed away for solstice 2013.

NOTES & REFERENCES

Chapter 1 notes (Skies And Earth)

01: Short unreferenced summary: Our world is some distance away from the galactic centre. We see the Milky way as a side-on view from within the disc of our galaxy. That disc has a polar axis at right angles to the disc: The North Galactic pole (NGP) is in Coma Berenices (near M64), with the other end going towards Sculptor. Due to precession, the stars rotate around the ecliptic pole (at $l = 96.38°$, $b = 29.81°$) causing the NGP to move exceptionally slowly, back and forth in the sky. A bit like a swaying branch. It would have come to its own "standstill", passing almost directly overhead, for an extended period during the Neolithic. Because the earth was decreasing its axial tilt over that period, that apparent "standstill" would have been for an extended period.
02: See https://en.wikipedia.org/wiki/Stonehenge_replicas_and_derivatives and https://clonehenge.com/
03: For examples see:
https://en.wikipedia.org/wiki/Geocentric_model
https://en.wikipedia.org/wiki/Socrates
https://en.wikipedia.org/wiki/Pre-Socratic_philosophy
04: 'A' Post-holes: Cleal *et al.* 1995: 142, 146, 269,281, 289, 290, 305, 467, 469

Chapter 2 Notes (An Introduction to Stonehenge)

<u>General introduction:</u>
01: Henges peculiarity to Britain: Parker Pearson et al, 2012: 322
02: Henges exclusive to Britain: Dyer 1990: 64
03: Uniqueness of carving at Stonehenge: Pitts, 2001: 85
04: Weight of Stones: Johnson 2008: 136-7
05: Mortice and tenon work was first noted in Camden's Britannia of 1586: Johnson 2008: 46
06: Diameter of sarsen circle: Chippendale, 2004: 19
07: Weight of Stones: Johnson 2008: 136-7

The people:
11: Genetics of people; past to present: Darvill, 2006: 70
12: Boscombe Down: Johnson 2008: 30
13: Amesbury Archer: Johnson 2008: 29
14: Amesbury Archer: Parker-Pearson 2012: 213
15: Evidence for trade with continent Bronze age: Pryor, 2004: 296-297
16: Genetics: Parker-Pearson 2012: 19
17: Clothing: Parker-Pearson 2012: 15
18: Landscape: Souden, 1997: 103
19: Granaries and storage: Parker-Pearson 2012: 25
20: Notes on stratification of society and elitist theory: Castleden 1987: 206-207
21: Rope use, houses and other materials: Parker-Pearson 2012: 16
22: 2300BC: Possible end date for long term use of Durrington using carbon dates: Parker-Pearson 2012: 117-118
23: Deer Antlers: Pitts, 2001: 186
24: Durrington feasting & construction phase: Parker-Pearson 2012: 110
25: Width and depth of river: Parker-Pearson 2012: 156
26: Boats 6-5000BC and Early Bronze Age: Pryor, 2004: 294
27: Wessex skulls: Castleden 1987: 194-195
28: Build & health of population: Castleden 1987: 195-198
29: Pig feasting & evidence of some summer occupation: Parker-Pearson 2012: 126
30: Stew & winter feasts: Burl, 2006: 164
31: Import of animals to Durrington: Parker-Pearson 2012: 120
32: Pig teeth & winter slaughter: Johnson 2008: 26

The path to Stonehenge:
41: Pre-Roman trails: Trevelyan, 1947: 8
42: Composite map compiled from: Wright, 1988: pages 10-30
43: Composite map compiled from: Trevelyan, 1947: 7-9
44: Composite map compiled from: Castleden, 1987: 115
45: South Downs National Park Authority. Website available at: http://www.nationaltrail.co.uk/southdowns/text.asp?PageId=30
46: South Downs track: Staveley, 2003: 3
47: Clearance: Butler, 2005: 2
48: The Sweet Track: Castleden, 1987: 118-119
49: Castleden, 1987: 115
50: Holes in former car park: Cleal *et al.* 1995: 43 & seq
51: Scheduled Monuments surrounding Stonehenge: Darvill, 2006: Preface (page 13)

Stonehenge: The word:
61: Hill, 2008: 22
62: Burl, 2006:225
63: Derivation of the word 'Henge': Mohen, 2002: 162
64: Kendrick: Pitts, 2001: 26
65: Stonehenge and the word 'henge': Pitts, 2001: 26 & 28
66: Meaning of Heng according to Stuckeley (Gallows): Burl, 2006: 19

A history of discovery
71: Deed of 937 AD: Hill, 2008: 21
72: Pepys Quote: Hill, 2008: 39
73: Defoe quote: Hill, 2008: 39
74: No Roman references: Hill, 2008: 39
75: Second wonder: Johnson 2008: 37
76: Historia Anglorum: Hill, 2008: 21
77: A digital scan of the work by Indigo Jones can be found at; http://digital.library.wisc.edu/1711.dl/DLDecArts.StoneHeng
78: John Aubrey & Walter Charleton: Johnson 2008: 55
79: Aubrey holes: Johnson 2008: 56-57
80: Ideas suggested by William Stukeley: Johnson 2008: 61-66
81: Ideas suggested by John Wood: Johnson 2008: 66 on
82: Stone numbering sequence: Johnson 2008: 123
83: Darwin's visit; Darvill, 2006: 45
84: Image from: Barclay, 1895: Plan I
85: Restoration of the central structure: Johnson 2008: 83 onwards
86: Gowland's suggestions; Darvill, 2006: 45
87: Image from: Barclay, 1895.
88: Richards, 2007: 160
89: Hawkins: Parker-Pearson 2012: 46 (but see also Hawkins reference)
90: Solstice: Hoyle 1977: 14
91: Sarmizegetusa (Grădiste): Hoyle 1977: 15
92: Solsticial alignment; other Great Henges; Parker-Pearson 2012: 342
93: Alignments and Ruggles: Parker-Pearson 2012: 48-49
94: Alignment: Johnson 2008: 176

Chapter 3 Notes (A Potted History And The Theories)

A potted history of the monument's phases
01: Settlement size: Parker-Pearson 2012: 4 and 92
02: Durrington/Stonehenge simultaneity: Parker-Pearson 2012: 5
03: Mesolithic post location: Burl, 2006: 223

04: The car park holes & woodland: Johnson 2008: 114
05: Post hole dating: Darvill, 2006: 62 & location plan on page 64
06: Mesolithic post-hole dating: Parker-Pearson 2012: 135
07: 'Tamed Wildwood': Darvill, 2006: 24
08: Vegetation landscape: Darvill, 2006: 69
09: Great and Lesser Cursus: Burl, 2006: 84
10: Cursus Period: Parker Pearson et al, 2012: 323
11: Uniqueness of Cursuses to Britain: Parker-Pearson 2012: 144
12: Dating the Scottish cursuses: Parker-Pearson 2012: 144
13: Newgrange: O'Kelly 1998: 21
14: Knowth: Eogan 1986: 178
15: Alignments: Hutton1996: 4
16: Chronology revision: Parker-Pearson 2007: 627
17: C14 decay: Hoyle 1977: 25-26
18: Climatic conditions in third millennium BC: Darvill, 2006: 93
19: Grasslands & woods: Parker-Pearson 2012: 164
20: Internal post-holes: Johnson 2008: 110-113
21: Periglacial stripes: Parker-Pearson 2012: 245

Potted history: Stage 1
31: The Bank: Johnson 2008: 100
32: The counter-scarp bank: Johnson 2008: 101
33 Calibration of ditch: Parker-Pearson 2012: 43
34: Holes beneath the Bank Johnson 2008: 100-101
35: Entrances though the bank: Johnson 2008: 102
36: The southern entrance though the bank: Johnson 2008: 104
37: Burl suggests these holes were of an earlier period 2006 edition because of the Moon alignment: Burl, 2006: 101
38: Stage 1: New Phasing: Parker Pearson et al, 2012: 309
39: Time-line: Parker-Pearson 2012: 7
40: Level of Aubrey Holes: Johnson 2008: 108
41: Aubrey Holes: Johnson 2008: 104-109
42: Entry through bank relocated: Dyer 1990: 66
43: Chalk Plaques: Parker-Pearson 2012: 227

Potted history: Stage 2 to 5
51: Q&R incompleteness: Parker-Pearson 2012: 169 and dating: 310
52: Q&R hole rings: Johnson 2008: 129-134
53: Lintels on Q&R settings: Johnson 2008: 132
54: Ring within R ring: Johnson 2008: 131 (see figure text)
55: Ditch around the Heelstone: Pitts, 2001: 139

56: Stone hole 97: Johnson 2008: 118 Note also similarity to North and south barrows: 120
57: Solar rise over Heelstone: Burl, 2006: 113
58: Stage 3: New Phasing: Parker Pearson et al, 2012: 310
59: Presence of sarsen below Avenue banks: Parker-Pearson 2012: 247-248
60: Image from: Barclay, 1895.
61: Quotation regarding Avenue: Parker Pearson et al, 2012: 242
62: Avenue bend (absence of hollows): Field 2012: 34-35
63: Stage 4: New Phasing: Parker Pearson et al, 2012: 311
64: Y&Z: Johnson 2008: 167
65: Y&Z holes: Dating and contents: Darvill, 2006: 164

The Stones: A potted summary
71: Tonnage of Stone: Richards, 2007: 207
72: Silica in sandstone: Richards, 2004: 5
73: Sarsen structure: Darvill, 2006: 131
74: Evelyn Quote: Burl, 2006: 9
75: Glaciers: Parker-Pearson 2012: 63
76: Hill, 2008: 13
77: Geoffrey of Monmouth's history was in Latin and dated 1136 and also known by the name *Historia Regum Britanniae*.
78: Sarcens buried: Parker-Pearson 2012: 294
79: Transport distance: Burl, 2006: 168
80: Sarsen shaping: Darvill, 2006: 131
81: Use of Mauls: Richards, 2007: 207
82: Weight of mauls: Pitts, 2001: 85
83: Stone preparation (evidence for): Parker-Pearson 2012: 42
84: Chippendale, 2004: 19
85: Height of trilithon stones: Johnson 2008: 143
86: Level top: Uprights different lengths: Burl, 2006: 168
87: Faces look inwards: Johnson 2008: 144
88: Sarsens worked more finely on inside faces: Hill, 2008: 44
89: Architectural device of entasis: Burl, 2006: 35
90: Level top of lintels: Chippendale, 2004: 19
91: Trilithon surfaces: Johnson 2008: 137
92: Uniqueness of Great Trilithon outer facing surface: Johnson 2008: 139
93: Height of trilithons: Johnson 2008: 136, *but note burl (p 176) has noted different dimensions (6.1, 6.5 and 7.3m respectively)*
94: Mortise and tenon joints: Darvill, 2006: 125
95: Position of stone 56: Johnson 2008: 240 & 244
96: Position of Stone 56: Parker-Pearson 2012: 256
97: Dating the Trilithons: Parker-Pearson 2012: 132

98: Foot of Stone: Pitts, 2001: 157
99: Stone 54 packing: Johnson 2008: 129
100: Outer bluestone circle diameter: Johnson 2008: 158
101: Bluestone variability below ground Cleal et al. 1995: 29
102: Outer bluestone height & spacing: Johnson 2008: 158
103: Bluestone outer ring not dressed Cleal et al. 1995: 27
104: Stone 36: lintel: Johnson 2008: 159
105: Inner bluestone height and spacing: Johnson 2008: 162
106: Inner horse-shoe grading: Burl, 2006: 179
107: Inner bluestone oval: Johnson 2008: 161
108: Best stones in inner circle: Burl, 2006: 177
109: Inner bluestone previous arrangements: Johnson 2008: 163

<u>The axis of the monument</u>
111: Post-holes at entry: Cleal et al. 1995: 142-143. see also 269
112: Bearing of Avenue: Burl, 2006: 189
113: Causeway post holes: Johnson 2008: 173
114: Stone hole numbering at entry: Cleal et al. 1995: 269
115: Chippendale, 2004: 124
116: Slaughter Stone: Johnson 2008: 151
117: Slaughter stone shaping: Burl, 2006: 187
118: D&E holes: Johnson 2008: 152
119: 97 & Heelstone: Cleal et al. 1995: 289
120: Sequence: Ditch around 96 intersecting 97: Pitts, 1992: 149
121: Stone 97 pit: Parker-Pearson 2012: 42
122: Parallelogram of Station Stones: Burl, 2006: 151
123: North and South Barrows: Johnson 2008: 148-150
124: Wyeth, 2001: 5
125: Observations by Ruggles: Alignment to the moon: Burl, 2006: 154

<u>Theories</u>
131: Theory quote Johnson 2008: 91
132: Burl, 1979: 200
133: Early theories: Johnson 2008: 58-59
134: Theories: Hancock, 1998: xiv
135: Temple to the goddess: Meaden, 1992: 166
136: Hawkins 1974: Preface
137: Errors of alignment in Hawkins: Hoyle 1977: 55
138: Astronomical alignments: Hoyle 1977: 65
139: No account in documented history: Hoyle 1977: 91
140: Moon and early Stonehenge: Darvill, 2006: 143
141: Alignment theories: Pitts, 2001: 227

Chapter 4 Notes (Mirror, Mirror)

<u>Metals and Stonehenge</u>
01: Copper in Turkey: Roberts 1980: 52
02: Mirrors; earliest use of: Prendergast 2003, 3
03: Smelting: Friede and Steel 1976: 466
04: for example see: https://en.wikipedia.org/wiki/Smelting or, for historical use, see: https://www.cornwall.gov.uk/environment-and-planning/conservation/world-heritage-site/delving-deeper/mining-processes/smelting/
05: Ease of production: Chirikure 2015: 20
06: Refer to Pastscape record 1220505 at: https://www.pastscape.org.uk/hob.aspx?hob_id=1220505
07: Penhallurick 1986: 183-4
08: Reflectance of polished tin: Golovashkin and G. P. Motulevich. 1964: 310-317
09: Properties of Tin Oxide: Baco, Chik, and Yassin. 2012: 61-72.
10: Polishing: Kogel, Trivedi, Barker and Krukowski, 2006: 153
11: Refer to Pastscape record 13822327 at: https://www.pastscape.org.uk/hob.aspx?hob_id=1382327
12: Refer to Chitterne local historian site at: http://www.chitterne.com/history/claypits.html
13: Traces of copper at footing of Stone 56: Chippendale, 2004: 168
14: Metal & copper at Durrington: Parker-Pearson 2012: 125
15: Metal trading networks Johnson 2008: 121
16: Tin tablet references from *Britain, or a Chorographicall description of ... England, Scotland and Ireland.(1637)*: Burl, 2006: 22.
Stuckeley's subsequent observations (quoted by Burl) can be found in more detail on page 31 of Stukeley, William. Stonehenge, a temple restor'd to the British druids. London: Printed for W. Innys and R. Manby, at the West End of St. Paul's, 1740.
Persistent Link http://nrs.harvard.edu/urn-3:FHCL:10937246
Repository Collection Development Department. Widener Library. HCL. Institution: Harvard University.
17: Dating the Copper Age: Ditch digging and Tree felling dated to approximately 2500BC: Parker-Pearson 2012: 123-126
18: Sources of metals: Gerrard, 2000: 14
19: Tin from the east Mediterranean and Lavant area: Berger, D et all. 2019 (internet open access)
20: Copper and tin: Johnson 2008: 24

Chapter 5 Notes (Stonehenge and The Hinge)

01: Scheduled monument details can be found at: http://www.pastscape.org.uk/
02: Schmidt-Kaler and W. Schlosser 1984: 183
03: Johnson 2008: 96

Chapter 6 Notes (A 'T' Shape device & The Cosmos)

01: Heathen Gods: North, R, 1997
02: The Rune Poem: Dickens, 1915
03: The lodestar: Sobecki, 2011
04: Arecibo Observatory in Arecibo, Puerto Rico.
05: Abbott *et al*, 2012: 59. Weight of Stone 54 = 29 tons (assumed to be UK tons but this is not stated in the report). By comparison a 14 metre, 300mm pine pole weighs approximately 0.55 UK tons.
06: Field et al, 2014
07: Abbott *et al*, 2012: 59.
08: Abbott *et al*, 2012: 21
09: Abbott *et al*, 2012: 21
10: Abbott *et al*, 2012: 51
11: Abbott *et al*, 2012: 21
12: Abbott *et al*, 2012: 23
13: Abbott *et al*, 2012: 52
14: Abbott *et al*, 2012: 23
15: Redrawn image of Early Bronze Age gold sun disc held at Wiltshire Museum (wiltshireheritagecollections.org.uk). Accession number: DZSWS:2015.6

Chapter 7 Notes (The Size Of The World)

01: Ceasar, J: Chapter 14: *"The Druids.... are said there to learn by heart a great number of verses; accordingly some remain in the course of training twenty years. Nor do they regard it lawful to commit these to writing, though in almost all other matters, in their public and private transactions, they use Greek characters."*
02: For background on Eratosthenes's method refer to: https://en.wikipedia.org/wiki/Eratosthenes#Measurement_of_the_Earth's_circumference
03: Al-burini, one of the greatest scholars of the medieval Islamic era (https://en.wikipedia.org/wiki/Al-Biruni). This requires a mountain and optical equipment capable of locating the horizon. A good description of this method can be found as follows:

NOTES & REFERENCES

https://en.wikipedia.org/wiki/History_of_geodesy#Biruni.
04: http://www.pastscape.org.uk/ Refer to monuments numbered: 1542578 and 1542588
05: Prosser & Raw 2000
06: http://www.pastscape.org.uk/ Refer to monuments numbered: Firle: 405762, Wilmington: 408729 and Bourne: 408485
07: http://www.pastscape.org.uk/ There are too many monuments to list each run of inter-visible monuments, but for the example sequential run from Well Combe to Bourne Hill, refer to monuments numbered: 1522748, 1522906, 470005, 470000, 1565877, 970619, 470015, 971106, 971117, 469974, 469977, 470067, 469965, 469940, 469962, 469931, 408547, 408524, 408507, 408535, 408521 and 408502.
08: Integer shown to make calculation simple. However, decimal and fractional maths could also be used.
09: Curwen 1929: 209
10: Another straight line of posts exist at what seems a random direction and location within the circle. See Staveley; 2003: 8

Chapter 8 Notes (Naysayers To A Round World)

01: Monument number 408442
02: Monument number 408729
03: Monument number 408448, 408606, 408532, 408726
04: Monument number 408711 (and also 408445)
05: Monument number 408826
06: Carpenter *et all*: 11-13
07: Monument number 408527
08: Monument numbers 408779 and 1571442 and 1571447
09: Monument number 469931
10: Monuments numbered 469974
11: Monument number 405762
12: Note that not all locations shown on image. Some example monuments at:
- Isle of Wight: Monument number 459799
- Swyre Head: Monument number 456525
- Joburg (France) monuments are adjacent to the local nuclear power plant

Chapter 9 Notes (Beginnings)

01: Parker Pearson *et al*: 2008
02: CADW reference PE298 (refer to cadw.gov.wales to search)

03: CADW reference PE301
04-07: All CADW reference PE300. Note that this record was under review at the time of writing.
08: Bevins *et al*, 2014: 189
09: Thom, 1962.
10: Chippendale, 2004: 75
11: Kernel density estimate of foot length distribution for European feet shows male feet at an average of about 270mmm and female feet at an average of just under 250mm. Jurca *et al*, 2019
12: Four Bronze Age bowl barrows on Tan Hill, listed by Grinsell as All Cannings 1, 1a, 12 and 15 (Monument no 216025), a further barrow listed as 1b: (Monument no 216028) and Two Bronze Age bowl barrows, listed by Grinsell as Stanton St Bernard 2 and 3, located on Milk Hill 10: (Monument number 216039).
13: Green, 1997: 263
14: Nash, 2020: 4

Chapter 10 Notes (Fear: Early Monuments)

01: Holmberg 2020: 7-38
Refer to: http://futhark-journal.com/rok/ for resources and links
02: O'Kelly 1998:124
03: O'Kelly 1998:fig 45
04: My thanks to the staff at Newgrange for checking these dimensions
05: Refer to https://en.wikipedia.org/wiki/Newgrange
06: Approximate alignment verified on site by author
07: Approximate alignment verified on site by author. For more information see Brennan, Martin, The Stars and the Stones: Ancient Art and Astronomy in Ireland - Thames and Hudson (1983)

Chapter 11 Notes (Explaining The Fear)

01: For more information on Lunistices, refer to González & Belmonte, 2019
02: Burl 2006: 101 and Cleal 1995: 143

Chapter 12 Notes (Folklore)

01: Aboriginal memories: Nunn and Reid 2015
02: Troyes, (c. 1160-1180)

03: Baillie, 2005: 208
04: Gordon, 1914:46

Chapter 13 Notes (The Makers' Mark)

01: Abbott *et al*, 2012: 28
02: Abbott *et al*, 2012: 29
03: Abbott *et al*, 2012: 31
04: Abbott *et al*, 2012: 34
05: Abbott *et al*, 2012: 53 and 54
06: Abbott *et al*, 2012: 54.
Images of the Arreton Down axes (Isle of Wight) can be seen at British Museum reference 00784612001 (https://www.bmimages.com/)
07: Madgwick et al. 2019:

Chapter 14 Notes (The Two)

01: The error generated by the mathematics of the simple slope method (to estimate the world's size) is 0.014% at 1:60 and 0.056% at 1:30. A more accurate method uses the Pythagorean theorem to generate a cosine. Additional errors, due to the curvature of the Earth, are generated if the mountains are a long distance apart.
02: If a slope of 1:30 were capable of being measured down to sea level, the mountain would need to be:
30 x 30 x 30 / 2: 13,500 human feet or 3,550 metres high.
03: For the moon to be used, a full, or nearly full, moon would have to set or rise during the night. Whilst this would work, it would not be practical because getting to the mountain, or getting down from it, would then have to be done in complete darkness.
04: This condition was checked using UK Ordnance Survey maps. The condition set is a) two mountain peaks very close to each other of which; b) one must be approximately 880 to metres high and; c) the other at least 10 metres higher and; d) the high one must look over the low one towards an unobstructed view to at least 100 kilometres of sea and e) the view must be able to view a sunset or sunrise. Although possible that the conditions exist elsewhere, only Pen-y-Fan appears to have these features.
05: Line of sight azimuths were generated by importing scaled Ordnance Survey maps into AutoCAD, generating 5 kilometre overlay arcs from Pen-y-Fan, with obscuring heights for each arc, and then manually checking the OS map for mountains or hills in excess of the calculated obscuring heights within

the zones generated for each arc.
06: These distances and level differences can be found by consulting Ordnance Survey maps.
07: Sunset and sunrise azimuths (clockwise angles from North) can be found from organisations such as www.suncalc.org.
08: See British Geological Society records for senni bedrock locations
09: Ixer et al 2020: 14

Chapter 15 Notes (The Universe)

01: Anaximander: Lawson 2004: 91
02: Ptolemy: Lawson 2004: 34
03: Public Domain work by Bartolomeu Velho in 1568. A copy is accessible at: http://en.wikipedia.org/wiki/File:Bartolomeu_Velho_1568.jpg
04: Aristarchus: Lawson 2004: 19-20
05: Philolaus: Lawson 2004: 32-33
06: Aristotle: Lawson 2004: 33
07: Public Domain work by Andreas Cellarius. Illustration of the Copernican system, from the Harmonia Macrocosmica (1660). A copy is accessible at: http://en.wikipedia.org/wiki/File:Heliocentric.jpg
08: Public Domain work by William Cuningham. known as the "Coelifer Atlas" from The Cosmographical Glasse of 1559. A similar copy is accessible at: http://www.loc.gov/exhibits/world/heavens.html
09: Johnson 2008: 101
11: If the winter founding point is initially set as being approximately one half radius (ie 8m for a 32m diameter bowl), so that platforms can be used for access to the end of the rod, then the height of the centre of the mirror-bowl above ground will be 8m x tan (51°) = 9.88m

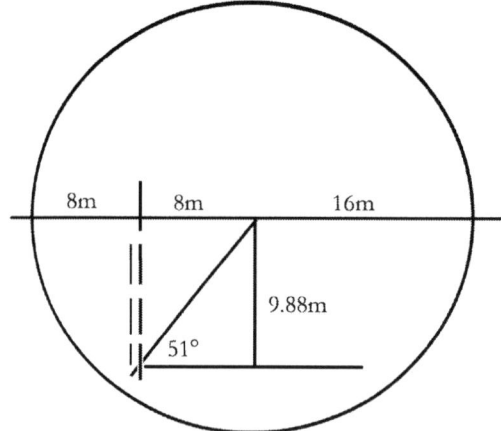

12: In winter at a 51° latitude, the maximum angle of the sun would have been about 15° in 2400BC (see appendix A). The rod will be pointing upwards by 24° relative to a perpendicular to the polar axis. The effect of this is that the rod must point *down* by 15° relative to the horizon (39°-24°). Because the apparatus is invertedly mimicking the sun, when the sun is up above the horizon by 15°, the rod points down by 15°:

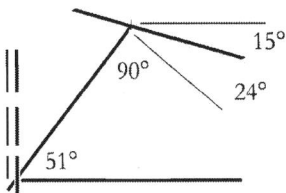

13: If the apparatus is pointing down by 15° over 9 metres, the centre of the mirror-bowl must be at least 9m x sin(15°) above the top of the rim: 2.33 metres. At a viewing angle of say up to 10° (see note below for explanation), an additional allowance of approx 9m x tan (10°) must be made because the end of the rod is set back from the rim: this adds 1.59 metres. Allowing for the size of the reflector adds a third of a metre. Another effect (in summer: see notes below) lowers the required maximum height by about 0.53 metres. In total, the height of the rim must be at least 4.8 metres or so below the mirror-bowl centre (2.33 + 1.59 + 0.33 + 0.53 = 4.78m).

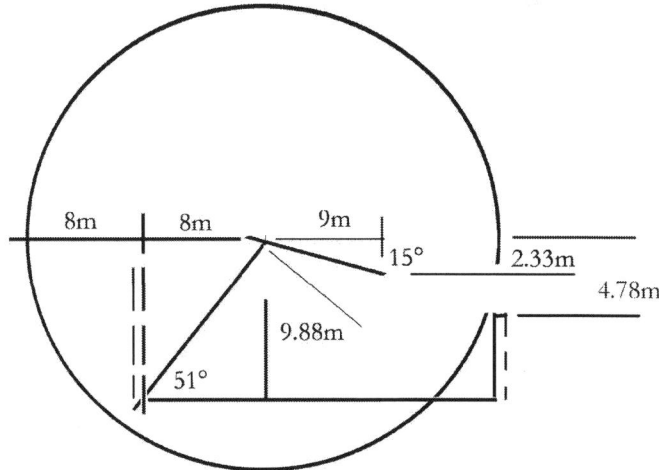

14: The viewing angle has been taken as up to 10°. This is an arbitrary selection because if people were made to stand, or say if children were excluded, the viewing angle could be a lot less than 10°. In this particular case, an angle of 10° seems appropriate because this equates to the start of the Avenue, just outside the bank of Stonehenge. Anyone allowed into the area within the bank could be able to see the effect from within the stones themselves; and anyone sitting around the bank would be able to see the same thing, even at its lowest point:

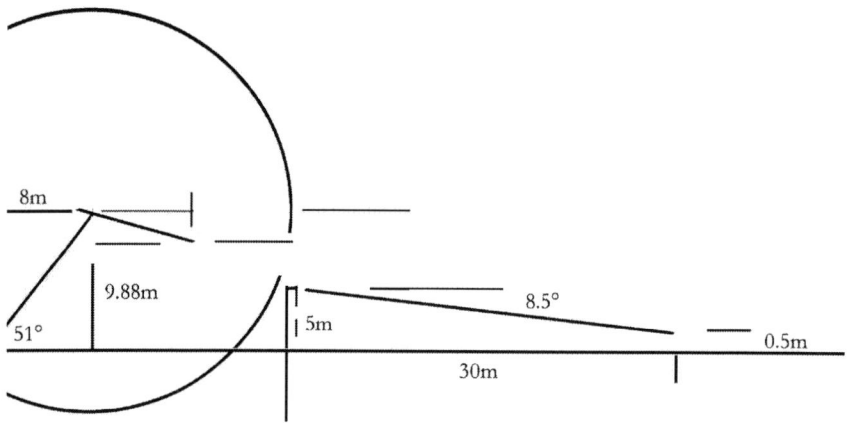

15: In summer at a 51° latitude, the maximum angle of the sun would have been about 15° in 2400BC (see appendix A). The rod will be pointing downwards by 24° relative to a perpendicular to the polar axis. The effect of this is that it will point down by 39°+24° = 63°. Because the apparatus is mimicking the sun, the rod must point down by 63° from the horizon's plane when the sun is up above the horizon by 63°:

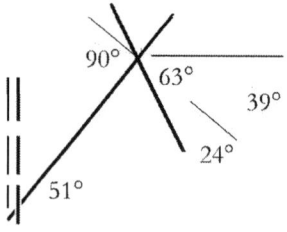

16: If the apparatus is pointing down by 63° in summer, this will extend the distance to the reflector when seen at an angle. The extended distance is 16m - 9m x cos (63°) = [16m - 4m] = 12m approx. For a viewing angle of 10°, This makes approx 12m x tan (10°) = 2.11 metres. Note that in the picture below, we are only concerned with the distance: the assembly also moves up in the summer months (see notes below)

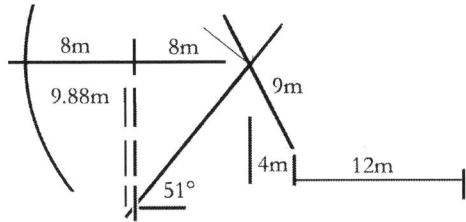

17: Given the parameters in the preceding notes, the height of the mirror-bowl rim *must not exceed* 9.88m -4.79m = 5.09m

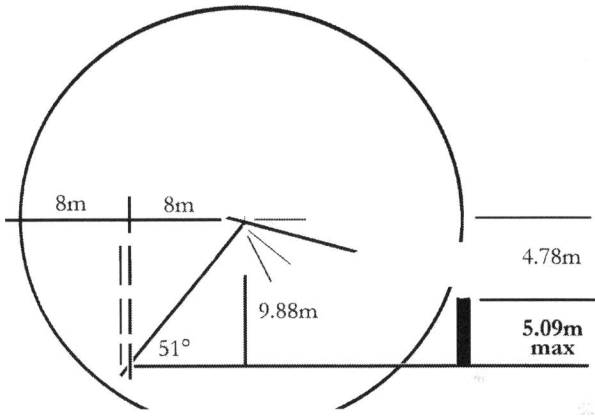

18: The inner rim diameter, for a 32m bowl diameter, must not exceed 30m because of the intersection with the rim (at a height of 5m or so):

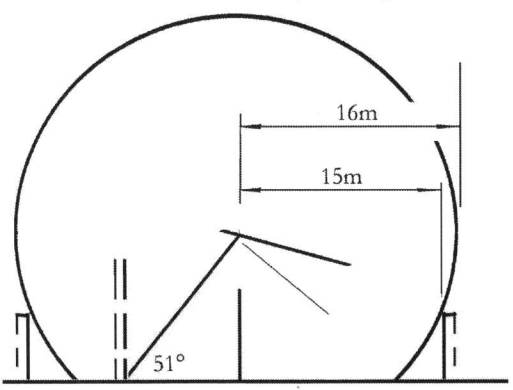

19: This note is calculated from the vertical range of the end of the sail when moved over the distance and multiplied by the difference between the polar position factors. In this case (8m + 1m) x (2 x sin (24°)) x cos (51°) x 90 % = 4.15m

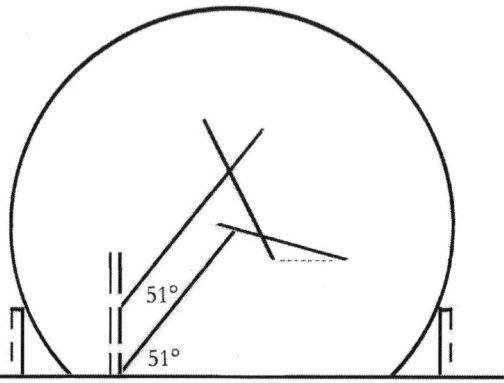

20: Using a pine post, 3-400mm diameter with a 14m length and allowing up to 1000kg weight, the lateral 'kick' force at the uppermost socket could be up to 10kN (for example if ropes are used for the final lift from say 30° to 51°). This results in an overturning moment at the base of up to 50kNm when lifting using the uppermost socket. The resistance of this (heaviest) stone is 284kN (see chap 2, note 5) multiplied by its lever arm in the short axis; >100kNm resistance.

21: The bowls for each position at Stonehenge are illustrated in this diagram:

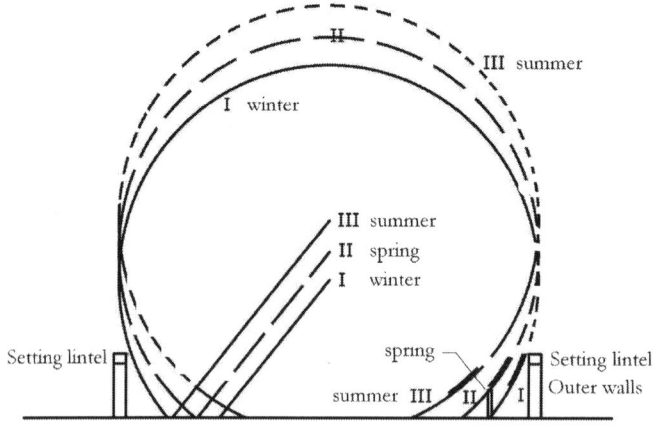

NOTES & REFERENCES 225

22: Illustration (showing the Great Trilithon & support for pole removed) to show possible drop-down for a timer-ring:

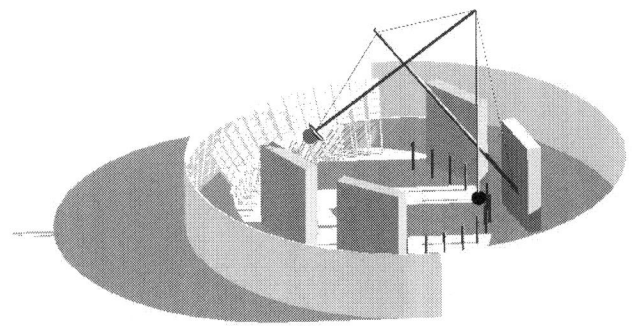

23: The Chief's Face was discovered by Terence Meaden in 1999 and reported by BBC news at:
http://news.bbc.co.uk/1/hi/sci/tech/474977.stm

24: A geocentric world is shown in more detail, locating where Salisbury and Stonehenge are positioned, in the illustration below:

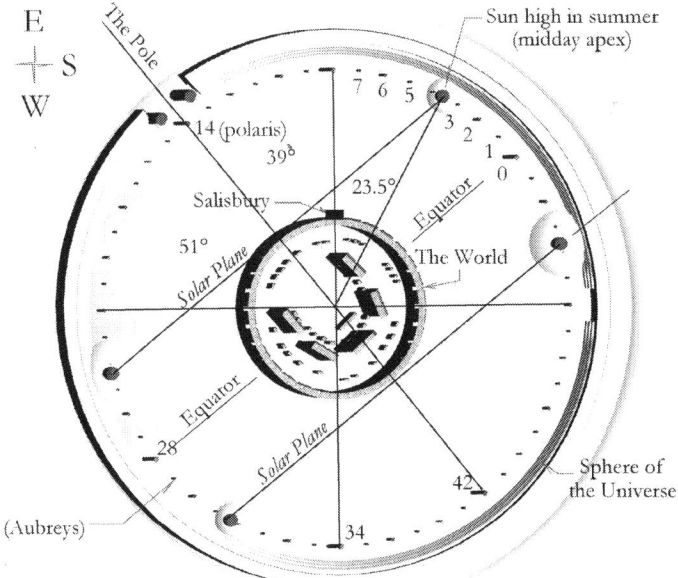

25: Hutton 1996: 6: *"Yule in Old Norse is Yol, Swedish jul and Danish juul. The derivation baffles linguists and is possibly related to the Gothic 'Huel' or Anglo-Saxon 'Hweal', meaning 'Wheel'. Another explanation is that it means 'Jolly'. "*
26: Diodorus Siculus, Library of History 5.67.06
27: Diodorus Siculus, Library of History 2.47.05

Chapter 16 Notes (Epilogue)
-

Notes for Appendices:

01: See: http://www-istp.gsfc.nasa.gov/stargaze/Sprecess.htm
02: Berger 1976 133. A copy may be accessed here: http://adsabs.harvard.edu/abs/1976A&A....51..127B
03: See: http://en.wikipedia.org/wiki/Solar_irradiation#Solar_constant
04: See: http://en.wikipedia.org/wiki/Airmass#Airmass_and_solar_energy
05: See: http://en.wikipedia.org/wiki/Ozone_layer
06: Page 41: 'Sustainable Energy – without the hot air' can currently (2013) be downloaded as a free .pdf at: http://www.withouthotair.com/

References

Abbott, M., Anderson-Whymark, H. *et al. Stonehenge Laser Scan: Archaeological Analysis Report 6457*. English Heritage, London, 2012

Baco, A. Chik, and F. Yassin. *Study on Optical Properties of Tin Oxide Thin Film at Different Annealing Temperature*. Journal of Science and Technology, 4(1), (2012).

Barclay, E. *Stonehenge and Its Earth-works*. D. Nutt, London, 1895

Berger, A. L *Obliquity and precession for the last 5 000 000 years*. Astronomy and Astrophysics, vol. 51, no. 1, Aug. 1976. p. 127-135.

Berger, D. Soles, J. S. Giumlia-Mair, A. R. Brügmann, G. Galili, E. Lockhoff, N. and Pernicka, E. *Isotope systematics and chemical composition of tin ingots from Mochlos (Crete) and other Late Bronze Age sites in the eastern Mediterranean Sea: An ultimate key to tin provenance?* Plos One, 2019 https://doi.org/10.1371/journal.pone.0218326

Burl, A. *Prehistoric Avebury*, Yale University Press, Ltd., London, 1979

Burl, A. *Stonehenge: A new History of the World's Greatest Stone Circle*. Constable & Robinson Ltd, London, 2006

Baillie, M. and P McCafferty, P. *The Celtic Gods*, Tempus Publishing Limited, Stroud, 2005

Balfour, M. *Stonehenge and its Mysteries*. Hutchinson and Co., London, 1983

R.E. Bevins et al., *Journal of Archaeological Science 42* (2014) 179e193

Butler, C. *An interim report on recent excavations at the Long Man, Wilmington, East Sussex*, 2005

Ceasar, J: *The Gallic Wars*. Translations by W. A. McDevitte and W. S. Bohn. Ref: http://classics.mit.edu//Caesar/gallic.html

Carpenter, E, Barber, M and Small, F. *Research Report series no 22-2013*, English Heritage, 2013

Castleden, R. *The Stonehenge People*, Routledge & Kegan Paul Ltd, London, 1987

Chippendale, C. *Stonehenge Complete*. Thames and Hudson Inc. New York, 2004

Chirikure, S. *Metals in Past Societies*. Springer, 2015

Cleal, R. M. J. Walker, K. E. and Montague, R. *Stonehenge in Its Landscape*. English Heritage, London, 1995

Curwen, E.C. *Neolithic Camp, Combe Hill, Jevington,* SAC Vol. 70 1929

Darvill, T: *Stonehenge: The Biography of a Landscape*. Tempus Publishing Limited, Stroud, 2006

Dickens, B: Runic and heroic poems of the old Teutonic peoples, ed. by Bruce Dickins. Published: Cambridge [Eng.] The University Press, 1915.

Dyer, J. *Ancient Britain*. B. T. Batsford Limited, London, 1990

Eogan, G. *Knowth and the passage-tombs of Ireland,* Thames and Hudson Ltd, London, 1986

Field et al, *The Avenue and Stonehenge: Archaeological Survey Report: Series no. 31-2012*, English Heritage, Portsmouth, 2012

Field et al, *Analytical Surveys of Stonehenge and its Immediate Environs, 2009–2013: Part 1 – the Landscape and Earthworks*. Proceedings of the Prehistoric Society, 80, pp 1-32, 2014 (doi:10.1017/ppr.2014.6)

Friede, H and Steel, R. *Tin mining and smelting in the Transvaal during the Iron Age*. Journal of the South African Institute of Mining and Metalurgy. July 1976

Gerrard, S. *The Early British Tin Industry*, Tempus Publishing Limited, Stroud, 2000

González-García, A. C and Juan A. Belmonte, J.A, *Lunar Standstills or Lunistices, Reality or Myth?* Journal of Skyscape Archaeology v5, no 2, p177-190, 2019.

Gordon, E. O. *Prehistoric London*. The Covenant Publishing Co., Ltd. 1914

Green, C. P. *The Provenance of Rocks used in the Construction of Stonehenge*. Proceedings of the British Academy, 92, 257-270. 1997

A. I. Golovashkin and G. P. Motulevich. *Optical and electrical properties of tin*, Sov. Phys. JETP 19, 310-317 (1964)

Hancock, G. and Faiia, S. *Heaven's Mirror: Quest for the Lost Civilisation*, The Penguin Group, London, 1998

Hawkins, G. S. *Stonehenge Decoded*. Souvenir Press Ltd, Great Britain, 1974

Hill, R.H *Stonehenge*. Profile Books, London, 2008.

Holmberg, Per *et al*. Futhark: International Journal of Runic Studies, ISSN 1892-0950, E-ISSN 2003-296X, Vol. 9-10, 2020.

Hoyle, F. *On Stonehenge*. Heinemann Educational Books, London, 1977

Hutton, R. *The Stations of The Sun*. Oxford University Press, Oxford, 1996

Ixer, R. A et al. *No provenance is better than wrong provenance': Milford Haven and the Stonehenge sandstones*. Wiltshire Archaeological & Natural History Magazine, vol. 113 (2020), pp. 1–15

Johnson, A. *Solving Stonehenge: The New Key to an Ancient Enigma*. Thames & Hudson, London, 2008

Jurca, A., Žabkar, J. & Džeroski, S. *Analysis of 1.2 million foot scans from North America, Europe and Asia*. Sci Rep 9, 19155 (2019). https://doi.org/10.1038/s41598-019-55432-z

Kogel, JE, Trivedi, N. C. Barker, J. M. Krukowski, S. T, *Industrial Minerals and Rocks; Commodities, Markets and Uses*. Society for Mining, Metallurgy, and Exploration, Inc (2006)

Lawson, R. M. *Science in the Ancient World: An Encyclopedia*. ABC-CLIO, California, 2004

Loomis, R.S. *Celtic Myth and Arthurian Romance*. Academy Chicago Publishers, 1997

Meaden, G. T. *The Stonehenge Solution*. Souvenir Press Ltd., London, 1992

Mohen, J-P. *Standing Stones: Stonehenge, Carnac and the World of Megaliths*. Thames and Hudson, London, 2002

Monmouth, G. *Historia Regum Britanniae (English: The History of the Kings of Britain)* c. 1136

Nash, D. J et al. *Origins of the sarsen megaliths at Stonehenge*. Sci. Adv. 2020; 6 : eabc0133. DOI: 10.1126/sciadv.abc0133, (2020)

Timothy Darvill[4], Susan Greaney[5], Georgios Maniatis[1], Katy A. Whitaker[6,7]

North, R. *Heathen Gods in Old English* UU Literature. Cambridge: Cambridge University Press, 1997

Nunn, P. D. & Reid, N.J. *Aboriginal Memories of Inundation of the Australian Coast Dating from More than 7000 Years Ago*, Australian Geographer, 2015 10.1080/00049182.2015.1077539

O'Kelly, M. K.*Newgrange*. Thames and Hudson, London 1982, 1998

Parker-Pearson, M. and The Stonehenge Riverside Project. *Stonehenge: Exploring the Greatest Stone Age Mystery*. Simon and Schuster, London, 2012

Parker-Pearson, M. et al. *The Age of Stonehenge*. Antiquity 81 (2007): 617-639

Parker-Pearson, M. et al. *Waun Mawn stone circle: the Welsh origins of Stonehenge Interim report of the 2018 season*. https://www.bluestonebrewing.co.uk/wp-content/uploads/2019/05/Waun-Mawn-2018-interim-report-lite.pdf: 2018

Penhallurick, R D. *Tin in Antiquity: Its Mining and Trade Throughout the Ancient World with Particular Reference to Cornwall*. Maney Publishing, 1986

Pitts, M. *Hengeworld*. Arrow Books, London, 2001

Prendergast, M. *Mirror, Mirror*. Basic Books, New York, 2003

Prosser, R. & Raw, M. *Landmark AS Geography*. Collins Educational, London, 2000

Pryor, F. *Britain BC*. Harper Perennial, London, 2004

Richards, J. *Stonehenge: The Story So Far*. English Heritage, Swindon, 2007.

Richards, J. *Stonehenge: A History in Photographs*. English Heritage, London, 2004.

Roberts, J. M. *The Pelican History of The World*. Penguin Books, Middlesex, UK, 1980

Schmidt-Kaler, T.H. and Schlosser, W. *Stone Age Burials as a hint to Prehistoric Astronomy*. Royal Astronomical Society of Canada, 78, 178. (1984) "Astronomie vor 5000"

Sobecki, Sebastian I. *The Sea and Englishness in the Middle Ages*. D.S. Brewer, Cambridge [978-1-84384-276-7] 2011

Souden, D. *Stonehenge: Mysteries of the Stones and Landscape*. Collins and Brown, London, 1997.

Staveley, D. *A Resistivity Survey of Combe Hill Causewayed Enclosure Near Willingdon, East Sussex*. 2003

Thom, Alexander. *The megalithic unit of length*, Journal of the Royal Statistical Society, A 125, 243-251, 1962.

Trevelyan, G.M. *History of England*. Longmans, Green and Co. Ltd., London, 1947

Troyes, C. *Perceval le Gallois, ou le Conte du Graal,* c. 1160-1180)

Wright, G.N. *Roads and Trackways of Wessex*. Moorland Publishing Co Ltd, Derbyshire, 1988

Wyeth, R. *The Stonehenge Story*. Gemini, Wiltshire, 2001

INDEX

24°; 45, 46, 51, 57, 68, 176, 190, 192
3D model; 204
51°; 45, 51, 52, 63, 175, 176, 178, 180, 193, 201
A posts; 6, 146
Advanced Research Projects Agency; 196
Africa; 27, 130
Air Force Cambridge Research Laboratory; 196
Air Mass; 191
Al-burini; 82, 83
Alderamin; 189
alignment; 5, 17, 19-24, 31-34, 55, 58, 59, 146, 187
alloy; 41
Alrai; 189
Altar Stone; 172
Amesbury Archer; 8
amphitheatre; 102
Anaraith; 33
Anaximander of Miletus; 173
Anglo-Saxon Rune Poem; 62
antler; 9, 21
Apollo; 186, 187
architecture; 28
Arecibo Observatory; 67, 196
Aristarchus of Samos; 174
Aristotle; 174
Arthurian; 149, 150, 186
astronomy; 17, 32, 33, 43, 120, 187
Atkinson, Richard ; 16
Aubrey holes; 14, 17, 22, 32, 55, 60, 175, 176
Aubrey, John; 14
Avebury; 3, 5, 10, 27, 33, 124, 125, 138, 143, 144, 147, 176
Avenue; 17, 22, 24-26, 55, 58-60, 68, 78, 175, 183
axial tilt; 4, 137, 190
Babylon Down; 94
bank; 12, 14, 21, 22, 25, 28, 31, 34, 37, 55, 60, 144, 175, 176
Barclay, Edgar; 15, 16, 25
barrow; 5, 25, 32, 39, 84, 86, 100, 111, 123-125, 138, 144, 146
Beachy Brow; 88-90, 108
Beachy Head; 48, 53, 84-86, 88, 91, 97, 100, 107
Beaker; 39
bearing; 31, 149
Bell Tout lighthouse; 84
Big Dipper; 189
bluestone; 23, 25-28, 30, 32, 71, 72, 118, 119, 125, 181, 182
Boscombe Down; 8
Bourne Hill; 48, 49, 86-92, 94, 96-100, 107, 111, 112, 121
Brecon Beacons; 113, 116, 165
bronze; 35
Bronze Age; 8, 9, 11, 40, 41
Burl, Aubrey; 24
Butser Hill; 11
Cæsar; 152, 154
Camden, William; 39
Camelot; 150
Cancer; 129, 130, 175
Canner, John Thomas; 39
Capricorn; 129, 175
carbon; 9, 18, 19, 21, 35, 37
cardinal; 22, 53-55, 58, 59
Carn Goedog; 112, 118, 119

INDEX

cassiterite; 35-38, 40, 42
Çatal Hüyük; 35
Cauldron of the Dagda; 152
chalk plaques; 22
Chief's Face, the; 182
Chilterns; 10
Chitterne; 39
Christchurch; 11
CIBSE Design Guide A; 141, 194
circle; 2, 7, 16, 22-24, 26-30, 32, 45, 51-54, 57-60, 67, 71, 78, 79, 93, 95, 96, 102, 120, 122, 127, 129, 130, 132, 133, 140, 141, 143-145, 153, 162, 175, 178, 180, 181, 189, 190, 192, 193, 195, 198, 199, 204
climate change; 126, 137, 200
clothing; 8
Coast; 3, 11, 47, 49, 110, 183
Cold Crouch; 94
Combe Hill; 11, 94-97, 99, 101, 114, 125
concentrator; 79, 150, 151, 179, 183, 184, 200
Copernicus; 174
copper; 8, 35, 39-42, 65, 68
Corn Du; 165-167, 169-171
Cornwall; 37, 38, 40, 42
Cosmos; 2, 50, 58, 60, 61, 140, 146, 147
Cotswolds; 10
counter-scarp bank; 21
Cove, the; 144
Craig Cerrig Gleisiad; 171
Cuningham, William ; 175
Curiosity; 40, 42
cursus; 14, 19, 20, 22, 187
Cwmcerwyn; 113, 117, 118, 161
D and E holes; 14, 24, 25, 31, 149
Darvill, Timothy ; 19
Darwin, Charles ; 15

De Bello Gallico; 152
De revolutionibus orbium coelestium; 174
De Situ Orbis; 152
De Troyes, Chrétien; 149, 151, 154
Defoe, Daniel; 14
Deneb; 189
Department of Energy and Climate Change; 200
dolerite; 23
Dowth; 131, 137
druid; 6, 14, 15, 33, 40, 81, 152-154
Drych Haul Cib Dâr; 6, 153
Durrington; 9, 11, 18, 39
earthwork; 2, 5, 11, 81, 144
Eastbourne; 11, 86, 108, 121
eclipses; 17, 33
Ecliptic; 145, 190, 192
Elliptic; 182, 190
enclosure; 7, 38
English Heritage; 76
engraving; 76
entasis; 28
entrance; 5, 15, 21-25, 139, 146
equator; 44-46, 51-53, 57, 63, 64, 128, 175, 176, 191-193, 203
equinox; 19, 20, 45, 48, 53, 68, 71, 74, 83, 92, 138, 139, 180, 181, 191-194, 199, 206
Etymology; 12, 13, 24, 184
Europe; 1, 7-9, 11-13, 18, 25, 40, 81, 124, 156, 157, 159, 160, 184, 186, 193
Evelyn, John ; 27
experiment; 3, 5, 35, 43-49, 86, 90, 92, 93, 97, 107, 117, 118, 132, 135, 183
farming; 19, 42
feasting; 9
Fir Bog; 151
Firle Beacon; 86, 100, 108, 110

Fisher King; 149
flake; 28
Foel Eryr; 112-115, 117, 123
Foel Feddau; 112-115, 117-119, 123
Folkington Hill; 99, 107, 111, 112
Fore Down hill; 108
Foxholes Brow; 88-90
France; 3, 38, 156, 159, 173
Galileo di Vincenzo Bonaulti de Galilei; 4
genetics; 8
geocentric; 2, 4, 5, 41, 50, 51, 56, 60, 63-65, 67, 72, 78, 80-82, 145, 148, 151, 152, 154, 155, 159, 160, 173-175, 177, 179, 183, 186, 187
Geoffrey of Monmouth; 27
Gerrard, Sandy; 40
glacier; 8, 27, 170
Gordon, William E.; 196
Gowland, William; 16, 39
graal; 149
grail; 6, 148-151, 154
granite; 27, 36, 40
Great Ridgeway; 10
Greek; 15, 24, 42, 152, 186, 189
Greenwich Observatory; 63
Guardians of the (North) Pole, the; 189
Halley; Edmond (or Edmund); 14
Harrow Way; 10
hauling; 179
Hawkins, Gerald ; 17, 33
Hawley, William; 16
haze; 47, 83, 150
hazel; 18, 19, 42
heel; 24, 25, 32, 59, 184
Heelstone; 24, 31, 32, 55, 58, 175
heliocentric; 4, 174
Helios; 24
Henge; 7, 12, 13, 17, 21, 31, 176, 184

Henry of Huntingdon; 14
Herstmonceux; 64
hinge; 12, 13, 25, 41, 43, 51, 154, 178, 179, 181, 184, 201
Historia Anglorum; 14
history; 6, 14, 19, 21, 23, 25, 26, 33, 82, 153, 154, 184
History of the Kings of Britain; 27
horizon; 1, 3, 14, 17, 20, 21, 46, 47, 49, 53, 54, 59, 63, 82-85, 87, 92, 105, 106, 113-116, 121, 124, 129, 131, 141, 142, 145, 161, 163, 165-167, 169
horse-shoe; 7, 26, 30, 72, 78, 179, 181, 182
Hoyle, Fred; 17, 33
hunter-gatherer; 8
hwēol; 24
Hyperborea; 186
Hyperion; 186
ice age; 8, 27
Icknield Way; 10
illusions; 54
Imbolc; 169
Indian Circle; 54
Ireland; 1, 3, 5, 8, 19, 20, 27, 151, 154, 156, 157
Jevington; 99
Johnson, Anthony; 17, 33, 34
Jones, Indigo; 14
Kendrick, Thomas; 12
King Charles the Second; 14
King James the First; 14
Knowth; 5, 19, 20, 22, 137, 138
Kochab; 189
Lake District, the; 8
landscape; 8, 19, 36, 37, 81, 123, 170, 172
language; 12, 13, 25, 62, 183
latitude; 32, 45, 52, 55, 57, 63, 130,

141, 146, 156, 176, 177, 191, 193, 200, 201
Leonardo da Vinci; 185
light-box; 133, 134
lintel; 7, 16, 23, 28-30, 67, 70, 78, 111, 134, 177, 178, 180, 181
Long Man, the; 101, 102, 108, 111
Loughcrew; 139
Low, Ward; 196
lozenge; 133, 134, 208
lunistice; 4, 6, 128, 142, 145
MacKay, David; 200
Marlborough; 27
maul; 28, 42
May Pole; 69
Meaden, Terence; 33, 77, 182
megalith; 7, 24, 28
metal; 2, 12, 35, 36, 38-42, 65, 68, 79, 127, 149, 151, 153, 183, 185
Mid Wilts Way; 123
migration; 8
Milk Hill; 123-125, 144
Milky Way; 1
mining; 10, 37, 42
mirror; 35, 39, 65-69, 71, 73, 79, 138, 148, 153, 177-181, 183, 185, 195-201, 203, 206, 208
mortice and tenon; 29
Newgrange; 5, 19, 20, 22, 126, 133-138, 145
Newton, Issac ; 14
North Downs; 10
North Pole; 42, 43, 45, 63, 144, 183, 189, 190
North Star; 51, 53, 58, 59, 63, 67, 70, 148, 151, 189
North, John; 17
Northern Hemisphere; 45, 131, 190, 191
number sequence of the stones; 15

oak; 19, 38, 153
Observatory; 17, 33, 34, 63, 67, 196
obsidian; 35
Old Red Sandstone; 170
Old Sarum; 10
orbit; 4, 42, 50, 51, 55-57, 126, 128, 129, 136, 145, 189-192, 198
oval; 26, 30
parabola; 195-197
Pashley Hill; 88, 90, 108
passage tomb; 19
patent; 61, 196
Pen-y-Fan; 165-171
Penhallurick, Roger; 37
people; 1-5, 8, 9, 12, 18, 24, 42, 68, 76, 80, 87, 95, 97, 120, 121, 126-128, 136, 137, 144, 147, 151, 153, 154, 159-161, 163, 177, 185-188, 200
Pepys, Samuel; 14
Perceval; 149
Petrie, Flinders; 15, 120
pewter; 40
Pherkab; 189
Philolaus; 174
Phoenice; 189
Phonecians; 33
pick-axes; 9, 21
pig teeth; 9
Piggott, Stuart; 20
Pilgrim's Way; 10
pine; 13, 18, 19, 87
Pitts, Mike; 24
Pliny; 153
plumb-bob; 53, 58, 59, 70
polar axis; 4, 43, 46, 48, 52, 55, 58, 63, 73, 101, 127, 175, 177, 178, 190, 193
Polaris; 62, 189
Pomponius Mela; 152
Preseli; 3, 27, 112, 116, 117, 124, 125

processions; 26
Pryor; Francis Manning Marlborough; 9
Ptolemy; 173, 174
pyramid; 41
Pythagoras; 153, 174
Q and R holes; 23, 26, 181
quartz; 27
Queen Boadicea; 33
radiation; 191, 197
radiator; 201
radio carbon; 9
Ragnarok; 126
reflectance; 38, 65
renewable energy; 41, 195, 196, 200
River Avon; 9
Roman; 11, 12, 14, 33, 42, 62, 120, 152
rope; 9, 79, 85, 87, 179, 206-208
Ruggles, Clive; 17, 32-34
sailing; 9
Salisbury; 10, 27, 52, 55, 204
Samhain; 169
Sanctuary, the; 124
Sarmizegetusa; 17
sarsen; 2, 7, 16, 23-30, 32, 39, 58, 60, 78, 122, 125, 162, 177, 178
Saxon; 12, 62
scanning; 16, 137, 158, 159, 183
Scottish Islands; 156-158
Senni; 170, 172
shadow; 54, 73, 130-139, 179, 191, 193
Silbury Hill; 124, 125
siliceous cement; 27
sinusoidal; 180, 192, 193
Slaughter-stone; 25, 31, 58
socket; 63, 67, 69, 70, 75, 178, 180-182
solar energy density; 191

solar plane; 175, 180, 190, 192, 193, 199
solar sweep; 199
solstice; 5, 7, 14, 17, 19-22, 24, 32, 34, 55, 126, 127, 129-133, 135-138, 140-142, 175, 176, 184, 187, 188, 192, 193, 208
South Downs; 3, 11, 82, 84, 86-88, 90, 91, 94, 97-99, 110, 114, 125, 144
Southern England; 43, 45, 49, 52, 64, 68
Spear of Lug; 152
sphere; 3, 4, 43, 45, 46, 48-52, 66, 67, 80, 102, 110, 174, 177-179, 195, 198
spherical; 66, 67, 71, 98, 161, 173, 179, 181, 186, 195-201
St Catherine's hill; 49
standstill; 4, 5, 128, 129, 142, 143, 145
Station Stones; 24, 32, 56, 60, 140, 141, 175, 176
Station Stones ; 25
statistics; 19
Stone 11; 78
Stone 21; 78
Stone 36; 30
Stone 4; 76
Stone 53; 75-77, 157, 159, 182
Stone 54; 30, 63, 69, 70, 75-78, 178, 180, 182, 204
Stone Age; 41
Stone hole 96; 31, 32
Stone hole 97; 24, 31, 32, 55
Stone of Fál; 152
Stonehenge in its Landscape; 16
Stonehenge Laser Scan: Archaeological Investigation Report; 76
Stukeley, William; 12, 14, 15, 25, 28, 40, 120

sunrise; 7, 17, 32, 48, 53, 83, 92, 94, 106, 109, 123, 129-131, 133, 138, 139, 161, 164
sunset; 20, 32, 53, 83, 92, 106, 109, 123, 129-131, 161, 164, 165, 169, 187
Sussex; 64, 82, 101
Sustainable energy without the hot air; 200
Sword of Light; 152
Swyre Head; 49, 110
symmetry; 17, 23, 31, 32, 76, 78
T shape; 61, 62, 76, 77, 159, 182, 183
tablet of tin; 40, 183
Tan Hill; 123-125, 144
The Cosmographical Glasse; 175
The three season device; 68, 74, 148, 159, 195
theories; 33, 34, 154, 174
Thuban; 43, 59, 62
tin; 12, 13, 35-42, 65, 68, 79, 183-185
tin pest; 65
Tir; 62
tons; 7, 27-29, 31
track-way; 10, 11, 88
trade; 8, 39, 40
transport; 27
tree-pole; 58, 148, 151, 178-181, 188
trilithon; 14, 16, 23, 25, 26, 29, 30, 39, 73, 74, 122, 179, 181, 182, 208
Tuatha Dé Danann; 6, 151, 152, 154
Universe; 4, 41-45, 51, 52, 55-57, 67, 80-82, 97, 102, 143, 147, 148, 152, 154, 173, 176, 177, 184, 186-188, 198
upright; 28-31, 181
Uther Pendragon; 33
Vega; 189
Victorians; 179
visitor's car park; 11, 18
Wales; 8, 27, 125, 144, 161, 165, 172
Wansdyke, the; 123, 124
Waun Mawn; 114, 125
Well Combe; 90
Wessex; 9, 17
Wessex skulls; 9
Willingdon Hill; 86, 107
Wilmington Hill; 100, 101, 107, 109, 111
Winchester; 10, 11
Windover Hill; 101, 102
Winn's oak; 38
Wood, John; 15
woodworking; 7
Y and Z holes; 26
Yule; 184, 188
zodiac; 175

List of illustrations

- Newgrange seen from Knowth, Ireland- p.5
- Stonehenge: As seen from north-east- p.7
- Composite map of routes - p.10
- Artistic impression of a Neolithic house- p.11
- Illustration by Indigo Jones- p.14
- Barclay's plan of 1895 (Plan I)- p.15
- Barclay's vision of Stonehenge restored- p.16
- Sunset- p.17
- Illustration of the car park posts- p.18
- Newgrange: Ireland - p.20
- Artistic view of the early henge along the 'solstice alignment'- p.21
- Artistic impression: The Great Cursus and Stonehenge seen to scale from above- p.22
- Stonehenge: North east elevation- p.23
- The Heelstone's new position as seen from along the centre-line- p.24
- The Avenue: Barclay's extract from Stuckeley- p.25
- The Y and Z holes - p.26
- The outer perimeter of Avebury today - p.27
- Stones of the outer circle drawn from slightly off-centre - p.28
- A lintel of one of the trilithons - p.29
- Bluestones within the inner monument - p.30
- The fallen Slaughter-stone- p.31
- observatory- p.33
- Tin streaming ground at Haytor, Dartmoor- p.36
- Streams leading to the River Lemon, Dartmoor- p.36
- Dry banks and trenches at tin streaming ground, Dartmoor- p.37
- Tracing stars using the North Pole - p.43
- The clock of the stars - p.44
- Following the stars - p.45
- Summer arrangement (angles)- p.46
- The World if a disc - p.47
- The effect of haze of horizons- p.47
- Aligning to sunrise (tripods)- p.48
- Aligning to sunrise using tripods- p.48
- Swyre Head Tumulus- p.49
- A drawing of apparent geocentric movement of the sun- p.50
- Drawing the Universe (with England at the top of the world)- p.51
- The 56 divisions of the heavens from Southern England- p.52
- Modified & updated version of Stonehenge 1845 plan- p.53
- The path of the setting sun - p.54
- Setting out the heavens at Salisbury - p.55
- The sun's orbits in a geocentric Universe - p.56
- Modified version of 1845 plan (unknown early centre)- p.56
- The Earth and solar markers- p.57
- Finding north- p.58
- The Avenue stones which were kept- p.59
- Computer model of the Stonehenge grounds- p.59
- T Rune- p.62
- Tyr by Lorenz Frølich- p.62
- To trace the summer sun (on a geocentric world)- p.63
- An equatorial telescope at

INDEX

- Herstmonceux- p.64
- Pole and stick arrangement to show the sun rotating- p.64
- Cast tin after cooling - p.65
- Cast tin after hand polishing- p.65
- Using mirrors to light up a model of a geocentric sun - p.65
- Adjusting the mirrors - p.66
- Using spherical mirrors to make a focal device - p.66
- spherical - p.67
- The setting ring (winter mirrors)- p.67
- Stone 54- p.67
- Summer: The sail pointing down - p.68
- High level position of the 3-season socket- p.69
- Raising the Maypole- p.69
- Setting the angle and location - p.70
- Stones 53 and 54- p.70
- The equinox support requirement- p.71
- The outer bluestone circle - p.71
- Oval or horse-shoe counterweight ring - p.72
- Stonehenge's inner bluestones - p.72
- The loading and rotation platforms- p.73
- Platforms for equinox and summer- p.74
- Rotation of the sail - p.74
- View from the north-east- p.75
- An example of the type of engraving found- p.76
- View of Stone 54 showing the 'T' shape - p.77
- Close-up on Stone 54- p.77
- A gold sun-disc- p.79
- Seeing the curvature of the Earth- p.83
- Using sticks as horizon sight-lines - p.83
- The highest hilltop ridge adjacent to Beachy Head - p.84
- The cliffs near Beachy Head- p.85
- Using a trough as a level - p.85
- Looking east from Beachy Head - p.86
- The hills to the north of Beachy Head - p.86
- Bourne Hill Tumulus - p.87
- Sight-lines (diagram)- p.87
- The route from Beachy Head to Bourne Hill- p.88
- Foxholes Brow bowl tumulus- p.89
- View of Bourne Hill from Foxholes Brow- p.89
- Beachy Brow bowl tumulus- p.89
- View of Bourne Hill & Foxholes Brow from Beachy Brow- p.89
- Sighting on a flat bowl - p.90
- Location of other tumuli (near Eastbourne)- p.91
- After sunrise looking west- p.92
- Bourne Hill: Sunrise at equinox - p.92
- The angles needed for slope approximation of the size of the world- p.93
- Combe Hill: Tumulus A - p.94
- Seeing the curvature of the Earth- p.94
- Combe Hill (adapted from Map by Curwen)- p.95
- Combe Hill: Showing how to calculate the curvature of the Earth- p.95
- Combe Hill: Bowl Tumulus B - p.96
- Bourne Hill seen from the semicircle at Combe- p.96
- The route beyond Bourne Hill (Eastbourne)- p.99
- Folkington Hill barrow- p.99
- Wilmington Hill barrow- p.100
- The view to Wilmington Hill from Holt Brow barrow- p.100
- The Long Man, East Sussex - p.101
- The Long Man's Amphitheatre- p.102
- Following the rotation of the Heavens on a northern slope - p.102

- A simple water level- p.103
- An extended water level- p.103
- Wrapping hair to make a surveyor's level- p.104
- Finding levels- p.104
- Removing cumulative errors- p.104
- Creating a levelling platform- p.105
- Measuring the angle down to the horizon using a timber block- p.105
- Calculating the slope down to the horizon- p.106
- Using coins: Beachy Head at sunset - p.106
- Folkington Hill brow Neolithic mound- p.107
- Willingdon Hill brow Neolithic mound- p.107
- Fore Down hill Neolithic mound- p.108
- Firle Beacon Neolithic mound- p.108
- Looking east to the sea from Wilmington Hill- p.109
- Looking west to the sea from Wilmington Hill - p.109
- Location of hills with east and west views over sea- p.110
- Swyre Head, Isle of Purbeck, Dorset- p.110
- The mounds of Folkington (foreground) and Bourne Hills (high background) - p.111
- A schematic drawing of the Preseli Mountains- p.112
- Foel Feddau seen from Foel Eryr- p.113
- Foel Eryr seen from Foel Feddau- p.113
- Waun Mawn- p.114
- Eryr and Feddau seen from Waun Mawn- p.114
- Foel Eryr's cairn- p.115
- Foel Feddau and its cairn- p.115
- The world if a disc- p.115

- Rotating a stone trough to position- p.116
- Foel Cwmcerwyn- p.117
- The mound atop Cwmcerwyn- p.117
- The small cairn of Cwmcerwyn- p.118
- The panorama from Cwmcerwyn- p.118
- The location of Carn Goedog- p.118
- Carn Goedog seen from Feddau- p.119
- Stone 68- p.119
- A view down the trough of stone 68- p.119
- Counting to 30 using two hands- p.120
- A hill height - p.121
- The angle down from a hill - p.121
- Stonehenge numbers- p.122
- A perspective on the location of Milk and Tan Hills- p.123
- Tan Hill and the Wansdyke- p.123
- Tan Hill seen from Milk Hill- p.124
- Milk Hill seen from Tan Hill- p.124
- Silbury Hill seen from West Kennett long barrow- p.125
- Newgrange: A blacked out chamber- p.126
- Modern metal spiral toy- p.127
- The sway of a swaying branch if the whole tree moves- p.128
- Measuring solstice using sticks- p.129
- Northern Hemisphere- p.129
- The line of sunset- p.130
- Capturing the angle of sunset using sticks and shadows- p.130
- Winter solstice shadow lines at sunset in the Northern Hemisphere- p.131
- Winter solstice shadow lines at sunrise in the Northern Hemisphere- p.131
- Shadows on a wall- p.132
- The black dot: Concentrating a shadow- p.132
- Shadow casts in a chamber- p.132
- Shadows and effects at a distance-

INDEX

- p.133
- The entry to the chamber at Newgrange- p.133
- Newgrange's light-box- p.134
- Intertwined spiral symbols at Newgrange- p.136
- West Kennet Long Barrow- p.138
- Inside West Kennet Long Barrow: (Stone surface at rear falling from left to right)- p.138
- Loughcrew Cairn T- p.139
- The passage into Cairn T- p.139
- Loughcrew backplate markers- p.139
- A geocentric heavens: The position of Winter Solstice- p.140
- Midday sun at solstice- p.141
- Circles showing the Sun's Circle at Solstice- p.141
- The Sun's angles at Solstice- p.142
- The Sun and Moon: Southern (sun) Solstice with major and minor (moon) lunistices- p.142
- Solstice and Standstills set in a geocentric Universe - p.143
- Avebury's layout (with East at top of map)- p.143
- Avebury's bank, ditch and outer stone circle - p.144
- The Cove at Avebury- p.144
- Lunistice variation when seen at a horizon set- p.145
- Posts set to find the major standstill of the Moon- p.146
- The blocking stone at West Kennett- p.146
- The village of Avebury- p.147
- Legend: The possible components of the Grail - p.149
- The stone correlated to Arthurian lore- p.149
- Legend: A shining cup- p.150
- The stone correlated to Arthurian grail lore- p.150
- Legend: Tuatha Dé Danann- p.151
- Spokes of the World: Image partly based on Google Earth software- p.156
- Zones of the World: Image partly based on Google Earth software- p.156
- The World imposed on a plan layout of Stonehenge- p.157
- Stone 3: Scottish Island groups.- p.158
- Stone 4: British and Irish groups.- p.158
- Stone 5: Continental European groups.- p.159
- Two mountains- p.163
- 1:60 mountains- p.164
- Line of sight over Corn Du from Pen-y-Fan- p.165
- Unobscured view azimuths from Pen-y-Fan- p.165
- Corn Du, and the gap between mountains to the Atlantic, seen from Pen-y-Fan- p.166
- The path from Corn Du to Pen-y-Fan- p.166
- The cairn on the summit of Corn Du- p.167
- The viewing line from the Twin Peaks- p.167
- Pen-y-Fan seen from Corn Du- p.168
- Zooming in on the back-lit horizon at a winter cross-quarter- p.168
- Ground folds just below summit of Pen-y-Fan- p.169
- Geology close to Pen-y-Fan- p.170
- Looking east from the side of Pen-y-Fan- p.170
- The two peaks seen from the cliffs of Craig Cerrig Gleisiad- p.171
- The cliffs- p.171
- Beneath the cliffs of Craig Cerrig Gleisiad- p.172
- The Ptolemaic geocentric model of the

- Universe in 1568 - p.173
- The Copernican system by Andreas Cellarius - p.174
- A geocentric view: Extract from 'The Cosmographical Glasse'- p.175
- Stonehenge 1845 Ground plan with solstice lines shown- p.176
- Measuring at Stonehenge - p.180
- Stone 54- p.182
- The Last Supper- p.185
- West Kennet: The blocking stone placed at the entry to the passage- p.188
- Precession - p.189
- The Earth circling the Sun - p.190
- Movement of the Sun seen from the Equator - p.192
- Solar Planes; movement of the Sun seen at temperate latitudes- p.193
- The wheels of the sun- p.194
- Solar Azimuths: Extract from CIBSE Design Guide A- p.194
- Extract from US patent no 4170985 - p.196
- Ray reflection lines from a spherical mirror - p.197
- Zooming in to the half-radius point - p.197
- Spherical collection zones for a fixed mirror set- p.198
- Equinoctial solar sweeps of spherical mirrors - p.199
- Seasonal solar sweeps - p.199
- Ideal mirror zones for winter (6 month) period - p.200
- One of the early second stage proof tests (2008) - p.201
- Stage 3 and 4 tests- p.202
- The roller assembly - p.202
- Constructing the mirrors- p.202
- Focusing the mirrors- p.202
- Preliminary scheme drawing extracts (6 no)- p.203
- Fitting the model into Stonehenge- p.204
- Stone 54 at Stonehenge- p.204
- Stage 5 tests: 2013 - p.205
- Setting up the tests as Stonehenge- p.205
- Test frame 6 from front: 2013- p.206
- Test frame 6: 2013 - p.206
- Stage 6 tests: Willingdon in June 2013 - p.207
- Stage 6 tests: Stay ropes seen from below, June 2013 - p.207
- Stay-rope trajectory for lowest set of the pole - p.208
- Back-reflections: Stage 6 tests: Willingdon, June 2013 - p.208
- Notes: The winter founding point - p.220
- Notes: Winter solar angles- p.221
- Defining the rim- p.221
- Notes: Viewing angles- p.222
- Notes: Summer solar angles- p.222
- Notes: Effect on viewing angle of summer set-up- p.223
- Notes: Summary of mirror bowl rim height- p.223
- Notes: Intersection with the rim- p.223
- Notes: Vertical range of the end of the sail- p.224
- Notes: The three season device- p.224
- Notes: Timer-ring- p.225
- Notes: Locating where Salisbury and Stonehenge are positioned- p.225

Printed in Poland
by Amazon Fulfillment
Poland Sp. z o.o., Wrocław